AQUARIAN QUERIES

AQUARIAN QUERIES

AVATARS, PROPHETS AND THE SACRED TEACHERS
BY ZEPHYRO SOLON

This book is a work of spiritual and historical exploration. While it draws upon real events, cultures, and philosophical traditions, some elements may be presented from the author's perspective for educational and reflective purposes. Any resemblance to actual persons or events is either coincidental or intentionally used to convey deeper meaning.

Published in the United States by Amazon Kindle Direct Publishing (KDP)
ISBN: 979-8-218-75604-8

For more information, visit:
www.aquarianqueries.com

About the Aquarian Queries Series

Aquarian Queries is a living series of books designed to awaken inquiry in the soul. Rather than offering conclusions, it models a way of seeing—one that honors both spiritual resonance and intellectual clarity. This series weaves ancient wisdom and modern science, myth and experience, into a unified conversation. It is rooted in the belief that we are transitioning into a new age of consciousness: one marked not by control, but by coherence; not by hierarchy, but by holistic integration. Each book stands on its own, but collectively they form a comprehensive framework for navigating the shift into the Age of Aquarius. Some scrolls explore sacred texts and spiritual traditions. Others delve into science, symbols, personal transformation, and forgotten histories. You are not asked to agree with everything. You are invited to inquire, reflect, and remember.

The questions are alive. And so are you.

Structure of the Aquarian Queries Series

The series is organized into thematic parts, each exploring a different facet of the human search for meaning:

Part 1: Foundation of the Inquiry

1.1 The Foundational Scroll
1.2 Solon's Story
1.3 Consciousness

Part 2: Science and the Soul

2.1 Quantum Mysticism
2.2 The Holographic Universe
2.3 Ancient Energy and Sacred Geometry
2.4 Time, Space, and the Multiverse
2.5 Intelligence in Nature
2.6 Bridging the Divide

Part 3: The Great Spiritual Traditions

3.1 Paths to the Divine
3.2 Hinduism
3.3 Buddhism
3.4 Jainism
3.5 Christianity
3.6 Judaism
3.7 Islam
3.8 Sikhism
3.9 Indigenous Spiritualities
3.10 The New Age Movement
3.11 Taoism

Part 4: Sacred Texts and Teachings

4.1 The Bhagavad Gita
4.2 The Yoga Sutras of Patanjali
4.3 The Upanishads
4.4 The Dhammapada
4.5 The Bible (Old and New Testaments)
4.6 The Tao Te Ching
4.7 The Guru Granth Sahib

... ✧ ...

Part 5 – Avatars, Prophets and Sacred Teachers

1.

JESUS OF NAZARETH

Teacher, Healer, and the Light of the World

-ZEPHYRO SOLON

CONTENTS

5.1.1
Historical Context: Judea Under Rome

To understand the life and significance of Jesus of Nazareth, we must first step into the world in which he lived—a world of empire, tension, and religious expectation.

In the first century CE, Judea was a small but volatile province under the dominion of the Roman Empire. Although geographically modest, it was spiritually immense, home to a people who carried an ancient covenant and a collective memory of liberation from bondage. For centuries, the Hebrew scriptures had spoken of divine justice, prophetic warnings, and the promise of a coming deliverer—the *anointed one*, or *Messiah*.

The Roman presence in Judea was not merely political—it was psychological, spiritual, and economic. Roman governors such as Pontius Pilate maintained order through military force and heavy taxation. While some Jewish elites collaborated with Roman rule to preserve their own status, many ordinary people lived under strain. Banditry, poverty, and messianic uprisings were not uncommon.

Religiously, Judea was diverse and divided. The Temple in Jerusalem stood as the spiritual heart of the Jewish world, but opinions differed sharply about what faithfulness to God looked like under occupation.

- The **Pharisees** emphasized strict adherence to the law and oral tradition.
- The **Sadducees**, associated with priestly and aristocratic circles, controlled the Temple and often cooperated with Rome.
- The **Essenes** withdrew to the desert, awaiting divine intervention and the purification of Israel.
- The **Zealots** called for violent resistance and revolt.

Into this landscape came a young teacher from Nazareth—an unassuming village in Galilee, far from the religious and political centers of power. He did not come with armies or scrolls of law. He came with stories, healing hands, and a message of a new kind of kingdom—one not built by swords or sustained by taxes, but rooted in the heart.

That such a figure would soon capture the attention of crowds, provoke the ire of authorities, and leave a mark on human consciousness that still reverberates today is both extraordinary and, in some ways, inevitable. The soil was dry. The longing was great. The people were watching for a sign.

Jesus entered history not in a vacuum, but in a moment ripe with spiritual hunger and socio-political pressure. His teachings would challenge not only Roman power but the assumptions of his own religious heritage. And while much about his life remains debated—his birth, his miracles, his resurrection—what cannot be denied is that something moved through that time and place that would echo across the ages.

5.1.2
The Life and Ministry of Jesus

The canonical Gospels offer glimpses—not exhaustive biographies—of the man called Jesus. What emerges is not a political leader or religious reformer in the conventional sense, but something subtler, deeper, and more enduring: a presence that unsettled systems, healed wounds, and awakened hearts.

He was born in obscurity, in Bethlehem according to tradition, to Mary and Joseph. The Gospel narratives surround his birth with signs and symbols—angels singing to shepherds, wise men following a star, Herod's fear of a newborn king. Yet after this dramatic entry, the story falls silent for decades.

When Jesus reemerges, it is as an adult seeking baptism from John, a fiery prophet calling people to repentance in the wilderness. According to the Gospels, as Jesus rises from the Jordan River, a voice from heaven affirms, "This is my beloved Son." The Spirit descends, and with it, a sense of divine mission.

He begins traveling through Galilee, teaching in synagogues, healing the sick, and announcing that "the kingdom of God is at hand." But unlike others who claimed divine authority, Jesus moved freely among the poor, the outcast, and the overlooked. He called fishermen, tax collectors, and zealots to follow him. He touched lepers, spoke with women, forgave sinners, and even dined with enemies of the people.

His words did not merely explain; they disrupted. "You have heard it said... but I say to you," he declared, challenging conventional interpretations of the law. He taught not just right action, but right intention—urging purity of heart, generosity without show, and prayer rooted in intimacy with the Divine.

Jesus' ministry was not only verbal. He healed physical ailments, calmed storms, multiplied food, and even, according to the Gospels, raised the dead. These acts were not spectacles for attention but signs—*semia*—of a deeper reality breaking into the world. He described this reality as the "kingdom of heaven," not as a place in the clouds, but as a realm that could be tasted now, through compassion, humility, and awakening.

He did not claim power in the way others did. When the crowds tried to make him king, he withdrew. When asked about taxes, he deferred to Caesar. When challenged about violence, he told his disciples to put away their swords. His mission was not to claim the throne of Israel, but to reveal a different kind of sovereignty: the reign of divine love in the human soul.

His teachings offended some and inspired others. He called out hypocrisy in the religious elite, warned of spiritual blindness, and offered paradoxes: the first shall be last, the greatest shall serve, and to save one's life, one must be willing to lose it.

As his following grew, so did the opposition. His refusal to conform to expectations—whether as Messiah, miracle worker, or rabbi—placed him on a collision course with both religious authorities and Roman power. And yet, through it all, he remained rooted in love.

He wept for Jerusalem. He forgave those who betrayed him. He prayed alone in gardens and stood silently before his accusers. His life was not a performance but a transmission—a living example of what it means to be fully human and fully surrendered to the divine.

In Jesus, many see the embodiment of spiritual paradox: strength in gentleness, authority in humility, divinity in humanity. He lived without wealth, traveled without armies, wrote no books, and died a condemned criminal. Yet his words echo still, not just in liturgy and theology, but in the questions we continue to ask.

5.1.3
The Kingdom of God: Within and Among

Of all the themes woven through the life and teachings of Jesus, none is more central—or more mysterious—than his proclamation of the "Kingdom of God." It was the core of his message, the destination of his parables, the vision that animated his movement. And yet, he never defined it in a rigid or systematic way.

Instead, he invited people to imagine it.

To the poor and marginalized, it was good news. To the powerful and religiously secure, it was often a threat. In a time when kings ruled by force and empires expanded through violence, Jesus spoke of a kingdom that arrived through humility, forgiveness, and inner transformation.

He compared it to things easily overlooked: a mustard seed, a hidden treasure, yeast in dough. It was not a political realm to be seized or a physical territory to be mapped. "My kingdom is not of this world," he would later say to Pilate. But neither was it distant or abstract. "The kingdom of God is within you," he told his listeners—and also "among you."

This dual phrasing has puzzled interpreters for centuries. Is the kingdom something internal, a shift in consciousness or moral alignment? Or is it a collective reality, revealed in communities of compassion and justice? Perhaps it is both: a kingdom that begins in the heart but cannot help but overflow into the world.

Jesus' kingdom turns hierarchy upside down. "Blessed are the poor in spirit," he says. "Blessed are the meek." He identifies greatness not with dominance, but with service. In his final days, he bends to wash the feet of his disciples—a teacher becoming the servant of all. This is the pattern of his kingdom: not ascending a ladder of power, but descending into love.

The phrase "kingdom of God" may be jarring to modern ears. It can sound patriarchal, triumphalist, or bound to a specific religion. But in its essence, it names a reality beyond all such categories: a field in which divine will flows through human lives, where justice and mercy kiss, and where the sacred is not delayed until the afterlife but discovered in this moment, in this breath.

In the world of Jesus' time, many expected the Messiah to bring a literal kingdom—to defeat Rome, restore Israel, and reign from Jerusalem. But Jesus offered no armies, no uprising, no golden age built by force. Instead, he gathered a band of ordinary people and asked them to live as if the kingdom had already come.

This was not escapism. It was resistance through transformation. The kingdom was not about escaping the world, but redeeming it. Not about punishing enemies, but loving them. Not about rules, but relationships.

In the centuries since, some have turned the kingdom into dogma or delayed it to a future apocalypse. Others have tried to bring it by force. But Jesus' original vision resists such narrowing. It remains as open and expansive as his parables—inviting us not to define the kingdom, but to live into it.

Queries for Reflection

- What does it mean to seek the kingdom of God today?
- Can a spiritual realm be lived out in the physical world?
- If the kingdom is within us, what blocks us from seeing it?
- How might our lives look if we believed the kingdom was already here?

5.1.4
Parables, Healings, and Teachings of Love

Jesus rarely taught in direct commands or abstract theology. Instead, he told stories.

His parables were layered, open-ended, and often unsettling. They did not offer simple answers, but invited deeper listening. "He who has ears to hear, let him hear," he would say—as though understanding required not just intellect, but inner attunement.

The parable of the **Good Samaritan** overturned social and religious expectations. A man is beaten, robbed, and left for dead. Two respected figures—a priest and a Levite—pass by without helping. It is the Samaritan, a member of a despised group, who stops, cares, and pays for the man's healing. "Go and do likewise," Jesus concludes. Compassion, not status, defines righteousness.

In the story of the **Prodigal Son**, a rebellious youth squanders his inheritance and returns home in shame—only to be met with a father's embrace and celebration. No lecture, no punishment. Just welcome. The story reveals the nature of divine love: extravagant, unearned, and always ready to restore.

Other parables are more cryptic: a merchant searching for a pearl, a farmer scattering seeds, brides awaiting a bridegroom. These speak to the mystery of spiritual awakening—its unpredictability, its demand for readiness, and its hidden value.

Beyond his stories, Jesus was known for his **healings**. He touched lepers, restored sight to the blind, made the lame walk, and—according to Gospel accounts—even raised the dead. These miracles were not performed for display. They were acts of compassion and signs that the kingdom he spoke of was not theoretical—it was breaking into the world through real transformation.

In many cases, the healing was more than physical. When he forgave sins, welcomed the excluded, or defended a woman caught in adultery, he was restoring people to wholeness—socially, emotionally, spiritually. "Your faith has made you well," he often said, suggesting that trust in divine goodness was as healing as any touch.

At the heart of Jesus' message was **love**—not as sentiment, but as a way of being. He was asked, "What is the greatest commandment?" He replied with two:

"You shall love the Lord your God with all your heart, and with all your soul, and with all your mind."

And, "You shall love your neighbor as yourself."

But he went further. "Love your enemies. Pray for those who persecute you." This teaching was not easy then, and it is not easy now. Yet it remains one of the most radical invitations in spiritual history.

He taught that love was the fulfillment of the law. That purity came not from ritual but from the heart. That the sacred was found not only in temples but in acts of mercy, generosity, and grace.

He warned against hypocrisy, particularly among the religious elite, calling them "whitewashed tombs"—beautiful on the outside, but hollow within. He urged his followers not to seek status or praise, but to serve in secret, give without expecting return, and pray not with many words but with sincerity.

In Jesus' teachings, we find a vision of life grounded in humility, truth, and fearless compassion. He was not merely offering advice; he was offering a path—a way of being in the world that reflected the presence of the divine here and now.

Queries for Reflection

- What do the parables reveal that logic alone cannot explain?
- How might healing—physical, emotional, or spiritual—be understood today?
- Can love truly be extended to enemies in a world of conflict and fear?

- What might it mean to live not *about* love, but *from* it?

5.1.5

Crucifixion and Resurrection: Death Transcended

The crucifixion of Jesus is, by all historical accounts, a brutal and humiliating death—reserved by the Roman Empire for rebels, criminals, and those who threatened public order. And yet, within the Christian tradition, it became the central symbol of redemption and hope. How could such violence give rise to a spiritual revolution?

To many of his followers, Jesus' execution was not only a miscarriage of justice but the shattering of their hopes. The man they had believed to be the Messiah—God's anointed one—had been arrested, tried, beaten, and nailed to a cross outside Jerusalem. The Gospels describe how he was mocked, stripped, and offered vinegar in place of water. And yet, even in his final hours, Jesus continued to teach. "Father, forgive them, for they know not what they do." A whisper of mercy amid agony.

For those who loved him, the crucifixion must have felt like failure. But something unexpected happened: the story did not end there.

According to early accounts, the tomb was found empty. Some reported visions. Others spoke of direct encounters—walking, talking, even sharing meals with the risen Jesus. These were not ghost stories. They carried a startling claim: that death had not had the final word.

The **resurrection**, whether understood literally, symbolically, or mystically, became the cornerstone of Christian belief. It signaled something new: that the apparent defeat of Jesus was actually his victory. That the cross, an instrument of torture, had been transformed into a tree of life.

But how is this transformation understood?

To many, Jesus' death is seen as **atonement**—a sacrificial offering for the sins of humanity. Others see it as the ultimate **act of solidarity**: God entering fully into human suffering, not to escape it but to redeem it from within. Still others understand the cross as a **mirror**—revealing the cruelty of systems and the innocence that still forgives.

The resurrection, then, becomes not just an event in history, but a **pattern** written into the cosmos. That life follows death. That the seed must fall to the ground to bear fruit. That love, when poured out fully, does not vanish—but returns transformed.

Mystics and contemplatives across centuries have found in the passion of Christ a doorway into spiritual death and rebirth. The cross becomes an inward reality—the surrender of ego, the letting go of control, the descent into unknowing. And the resurrection becomes the emergence of a new self, rooted in trust, grace, and a deeper knowing of the divine.

From a Queries perspective, this sequence—death, silence, and return—is archetypal. It is echoed in myth and in nature, in spiritual awakening and personal loss. It is the shape of transformation.

Jesus did not avoid suffering. He entered it. He did not escape death. He transcended it. And in doing so, he offered not just a promise of life after death, but a deeper truth: that love, courage, and surrender have power even in the darkest hour.

Queries for Reflection

- Is the crucifixion a tragedy, a triumph, or both?
- What does resurrection mean beyond belief—what does it ask of us?
- Can suffering be a doorway to transformation rather than a dead end?
- What systems today would crucify love, and how do we respond?

5.1.6
Christianity Through the Ages

In the wake of Jesus' life, death, and reported resurrection, something unexpected happened. A small and persecuted movement began to grow—not by the sword, but through testimony, transformation, and a vision of divine love that could not be silenced.

The earliest followers of Jesus did not call themselves Christians. They were Jews who believed that Jesus was the long-awaited Messiah and that his life fulfilled the promises of the Hebrew scriptures. They met in homes, broke bread, and shared what they had. They spoke of him not just in memory, but as a living presence. They called him *Lord*—a title both spiritual and political, subverting the empire's claim that Caesar alone held that name.

As the message spread beyond Judea into the broader Roman world, it began to take new forms. Gentile converts brought different perspectives. Disputes arose. Letters were written. Communities formed. In time, the message of Jesus grew from a movement to a religion.

By the fourth century, the Roman Empire—once the executor of Jesus—would embrace Christianity as its official faith. With Emperor Constantine's conversion, the church gained legal status, followed by power, wealth, and institutional authority. The cross, once a symbol of execution, became a symbol of empire.

This transformation was both a victory and a tension. Christianity gained protection and permanence—but it also became entwined with politics and hierarchy. Councils were convened to define orthodoxy. Texts were canonized. Heresies were condemned. The spiritual revolution of Jesus was now administered by bishops, governed by doctrine, and protected by emperors.

And yet, through it all, a deeper current flowed.

Monastic communities emerged, preserving sacred texts and cultivating lives of prayer and service. Mystics spoke of union with the divine. Reformers called the church back to the heart of the Gospel. Artists, musicians, and poets gave expression to Christ's presence in beauty and symbol. And ordinary people found in the story of Jesus a source of comfort, guidance, and inspiration.

Over centuries, Christianity would fracture and reform.

- The **Great Schism** of 1054 divided Eastern Orthodoxy and Western Catholicism.
- The **Protestant Reformation** of the 16th century, led by figures like Martin Luther, challenged church corruption and returned focus to scripture and grace.
- New denominations would arise, each interpreting Jesus' message in its own light—some emphasizing faith, others works; some tradition, others experience.

Christianity would spread across continents—sometimes through mission, sometimes through conquest. In many cases, it blended with local cultures, creating new expressions and spiritual hybrids. In others, it clashed, leading to oppression, resistance, and revival.

In recent centuries, modernity brought new questions. Science, pluralism, and historical criticism challenged literal interpretations of scripture. Secularism and disillusionment led many to leave the church. Others sought to rediscover Jesus outside of dogma—through mysticism, social justice, or personal experience.

Today, Christianity remains the world's largest religion, yet also one of the most diverse and debated. For some, it is a living faith. For others, a wounded institution. For many, it is both.

What has endured is not always the form, but the essence: the call to love God and neighbor, to forgive, to serve, to seek the kingdom within. Beneath layers of tradition and complexity, the spark of Jesus still flickers—sometimes in grand cathedrals, sometimes in prison cells, sometimes in quiet hearts who still ask: Who was this man, and what does his life mean for mine?

Queries for Reflection

- How has the institutionalization of Christianity shaped its message?
- What have we gained—and lost—in translating a movement into a religion?
- Can the essence of Jesus' teaching be recovered beneath centuries of interpretation?
- What role does Christianity play in today's spiritual landscape?

5.1.7
Mystical and Esoteric Views of Jesus

Beyond the creeds and institutions of Christianity lies a more hidden thread—a mystical current that has quietly flowed through centuries, seeking not merely to worship Jesus, but to know Christ as a living presence within.

In these mystical and esoteric traditions, Jesus is not only a historical figure or savior, but a symbol of inner transformation. His life becomes a mirror of the soul's journey—from birth in the humble manger of the heart, through trials and surrender, into union with the divine.

Early Gnostic Christianity

Some of the earliest alternative views of Jesus emerged among the **Gnostics**, a diverse and often misunderstood movement in the first few centuries CE. Gnostic texts such as the *Gospel of Thomas*, *Gospel of Philip*, and *Gospel of Truth* emphasize inner knowledge—*gnosis*—as the key to salvation.

In these writings, Jesus is portrayed not as a sacrificial lamb, but as a revealer of forgotten truth. He speaks in riddles and paradoxes, urging seekers to wake up to the divine spark already within them. Salvation, in this context, is not escape from sin but awakening from ignorance.

From the *Gospel of Thomas*:

"If you bring forth what is within you, what you bring forth will save you. If you do not bring forth what is within you, what you do not bring forth will destroy you."

These teachings were eventually declared heretical by the institutional church, but their resonance can still be felt among modern seekers who view Jesus as a mystic guide rather than a distant deity.

Christian Mysticism

Throughout Christian history, mystics have continued to pursue direct union with the divine—sometimes at great personal risk. Figures like **Meister Eckhart**, **Teresa of Ávila**, **John of the Cross**, and the anonymous author of *The Cloud of Unknowing* spoke of an experience of Christ that went beyond words or images.

For them, Christ was not just a historical figure but an **inner presence**, born in the soul through contemplation and surrender. In their writings, silence becomes a sacrament, and love becomes the path.

Eckhart wrote:
"The same eye with which I see God, is the eye with which God sees me."

This non-dual vision suggests that Christ is not simply out there, but also in here—that the divine and the human are not separate, but mysteriously one.

Esoteric and Symbolic Views

Other traditions view the story of Jesus as an **allegorical template** for spiritual awakening. In the language of alchemy, the crucifixion becomes the dissolution of the ego; the resurrection, the rebirth of the true self. The manger becomes the humble heart; the cross, the place where human limitation yields to divine love.

In some esoteric interpretations, Jesus is an embodiment of the **Christ consciousness**—a universal presence that can awaken in anyone, regardless of religion. In this view, "the Christ" is not Jesus' last name, but a title for one who has fully realized the divine nature within and lives from that center.

This view does not diminish Jesus, but expands him—offering a way to see his life not only as exceptional, but exemplary.

A Quiet Thread

Mystical and esoteric understandings have always remained a minority voice—quiet, often hidden, sometimes suppressed. But they have never disappeared. In every age, individuals have looked beyond dogma to experience Jesus not merely as an object of worship, but as a doorway into the sacred depths of being.

For some, this approach feels liberating. For others, it may seem unsettling. But within the Aquarian spirit of inquiry, these perspectives invite deeper questions—not to replace the Jesus of faith, but to encounter him anew.

Queries for Reflection

- What might it mean to experience Christ, rather than only believe in him?
- Can the story of Jesus be a symbolic map of inner transformation?
- How do mystical perspectives shift our understanding of salvation?
- Is Christ a person, a presence, a process—or something beyond all these?

5.1.8
Jesus in Other Traditions

Though Jesus of Nazareth is most often associated with Christianity, his presence has reverberated far beyond its walls. In other world traditions—sometimes as prophet, sometimes as sage, sometimes as mystery—Jesus appears in unexpected ways, bridging divides and opening new avenues of understanding.

In Islam

In the Qur'an, Jesus is known as **ʿĪsā**, one of the most revered prophets in Islam. He is described as born of the Virgin Mary (*Maryam*), as a miracle-worker who healed the blind and raised the dead by the permission of God, and as the **Messiah**—though not in the Christian sense of divine sonship.

Muslims do not view Jesus as God incarnate, nor believe in his crucifixion in the same way. According to most Islamic interpretations, he was not killed but was raised by God, and will return at the end of time to restore justice. Despite these doctrinal differences, the reverence is profound. He is considered among the **Ulū al-ʿAzm**—the five great prophets, alongside Noah, Abraham, Moses, and Muhammad.

In Sufism, the mystical dimension of Islam, Jesus is often portrayed as the embodiment of spiritual purity and detachment. Rumi, the great Persian poet, wrote of Jesus as a soul untouched by the material world, a mirror of divine breath.

In Judaism

Within Judaism, Jesus is generally regarded as a Jewish teacher or reformer whose followers later diverged into a new religion. While Jewish tradition does not accept him as the Messiah, some scholars and mystics see Jesus as part of a long lineage of **prophetic voices** within Israel's tradition.

Kabbalistic thought, with its rich symbolism of divine emanations and messianic longing, offers conceptual bridges that resonate with the mystical portrayal of Jesus found in early Christianity. Some Jewish thinkers—especially in modern interfaith contexts—have sought to reclaim Jesus as a **Jewish mystic**, misunderstood by history but rooted in Jewish spiritual soil.

In Hinduism and Buddhism

In Hinduism, Jesus is sometimes viewed as a **divine incarnation** or *avatar*—a being who embodies God's presence on Earth. Figures like **Swami Vivekananda** and **Paramahansa Yogananda** spoke of Jesus with deep reverence, often describing him as a yogi—one who had achieved full union with the Divine.

The *Autobiography of a Yogi* describes Jesus as part of a spiritual lineage, a bridge between East and West. In this vision, Jesus is not an outsider, but a cosmic teacher whose realization of Christ-consciousness mirrors the awakened states described in the Vedic tradition.

In certain strands of Buddhism, particularly Mahāyāna, Jesus is likened to a **bodhisattva**—one who renounces nirvana to help all beings awaken. His compassion, his willingness to suffer for others, and his inner realization of non-attachment all echo themes found in the Buddhist path.

In Indigenous and Other Traditions

In some Indigenous spiritualities, Jesus is embraced as a **great spirit**, a wisdom carrier, or even an elder brother. His emphasis on peace, harmony with creation, and justice for the marginalized aligns closely with earth-based teachings that honor the sacred in all life.

Newer movements such as the Bahá'í Faith recognize Jesus as one in a line of progressive **Manifestations of God**, each revealing truth in different times and cultures. In this view, he is neither exclusive nor isolated, but part of an unfolding tapestry of divine guidance.

A Universal Presence?

These diverse portrayals raise a deeper possibility: that the essence of Jesus transcends religious boundaries. While doctrines vary, the qualities associated with him—love, forgiveness, healing, self-sacrifice, and divine union—speak to something universal in the human quest.

He becomes not only a figure of one tradition, but a mirror in many—each seeing in him a glimpse of its own highest ideals.

This is not to blur distinctions or erase theological nuance. Rather, it is to notice a mystery: that across continents and centuries, hearts have recognized in Jesus a presence that awakens awe, humility, and longing for the sacred.

Queries for Reflection

- What does it mean that Jesus is honored across so many traditions?
- Can multiple portrayals be true in different ways?
- Is it possible to understand Jesus more fully by seeing him through the eyes of others?
- What does a universal Christ suggest about the nature of spiritual truth?

5.1.9
Contemplating Christ: A Living Question

To study the life of Jesus is to engage more than history. It is to step into a living question—one that refuses to be reduced to dogma or dismissed as myth. His presence lingers, not just in scripture and sacrament, but in the quiet moments of conscience, compassion, and contemplation.

Who was this man?

To some, he is the Son of God, the Savior, the only way to eternal life. To others, a prophet, a mystic, a teacher of love and liberation. Some see him as a revolutionary, challenging the political and religious powers of his day. Others as a mirror of the divine potential within all human beings.

Each view says more than facts alone can verify. It reveals something of the lens through which we see the sacred.

Perhaps that is part of the mystery. Jesus does not impose an image of himself; he invites discovery. "Who do you say that I am?" he asked his disciples—a question that continues to echo, not only in churches but in the searching heart.

To contemplate Christ is to wrestle with paradox:

- A teacher who taught with authority, but knelt to wash feet.
- A man who forgave enemies while hanging on a cross.
- A presence claimed by religion, yet often at odds with its power.
- A being seen as both fully divine and fully human.

It is not necessary to resolve these tensions. In the spirit of *Aquarian Queries*, it may be more fruitful to sit with them—to let the figure of Jesus speak in layered ways across time, tradition, and personal reflection.

For some, he awakens devotion. For others, curiosity. For many, he becomes a bridge—between head and heart, matter and spirit, the human and the divine.

In a world still marked by injustice, fear, and fragmentation, the example of Jesus—his courage, his compassion, his inner clarity—offers not only consolation, but challenge.

He does not only ask to be believed in. He asks to be followed.

Not in mimicry, but in essence. Not by recreating his life, but by embodying the spirit he revealed—love without condition, truth without violence, and faith that acts in service of others.

In this light, Christ remains not merely a figure of the past, but a living presence—a question that continues to shape the journey of those who seek what lies beyond the visible.

Final Queries for 5.1

- Who is Jesus to you—not according to tradition, but in your own experience?
- Is Christ a name, a role, or a presence?
- What does it mean to follow a figure who walked without wealth, without force, and without fear?
- How might his message be lived anew in this age?

2.
THE BUDDHA

Awakener of the Path Beyond Suffering

-ZEPHYRO SOLON

CONTENTS

5.2.1
Siddhartha's World: A Life of Comfort and Questions

Long before he became known as *the Buddha*—the Awakened One—he was **Siddhartha Gautama**, born into a world of luxury and lineage, yet destined for a path of radical renunciation.

According to tradition, Siddhartha was born in the 6th or 5th century BCE in what is now Nepal, into the ruling warrior caste of the Shakya clan. His father, a regional king or chieftain, is said to have shielded him from all suffering, surrounding the young prince with beauty, pleasure, and comfort. Lavish palaces, fine clothing, music, and delicacies were his everyday reality.

But even in this curated paradise, something stirred.

The early legends tell of Siddhartha's encounter with what have come to be called the **Four Sights**: an old man, a sick person, a corpse, and a wandering ascetic. These were his first glimpses of the human condition beyond palace walls—of aging, illness, death, and the search for something deeper than worldly pleasure.

These sights cracked the illusion of permanence and control. For the first time, Siddhartha saw that no wealth, privilege, or protection could prevent the inevitabilities of life. His questions deepened:

What is the cause of suffering?

Is there a way beyond it?

Can peace be found that is not subject to loss?

His world of pleasure, once satisfying, now felt hollow. The comforts of the palace could no longer mask the existential ache.

At the age of 29, Siddhartha made a momentous decision—one that echoes through history as **The Great Renunciation**. Leaving behind his wife, newborn son, and royal status, he set off into the forests of India in search of truth.

This act was not a rejection of life, but of illusion. He was not fleeing responsibility, but seeking understanding. Like Jesus retreating to the wilderness or Moses climbing Sinai, Siddhartha withdrew from the world not to escape it, but to comprehend it.

He entered a world of wandering sages, yogis, and seekers. Ancient India was a landscape rich with spiritual experimentation. The Vedas had long spoken of cosmic order, sacrifice, and liberation. The Upanishads had begun to whisper of a Self beyond self—*Atman* and *Brahman* and the mystery of non-duality.

But Siddhartha was not content with inherited answers. His path would be one of direct insight.

In Siddhartha's world, the divide between wealth and suffering, certainty and questioning, was stark. He had lived at both poles. And in that contrast, a seed was planted—not for religion in the institutional sense, but for **a way of seeing**.

Before he became a teacher, he was a questioner.

And perhaps that is where all awakening begins—not with answers, but with the courage to ask.

Queries for Reflection

- What illusions of comfort might shield us from deeper truths?
- Can the presence of suffering open the door to spiritual insight?
- What does it mean to leave behind certainty in search of understanding?
- Is renunciation an act of loss, or of alignment?

5.2.2
The Great Renunciation and Search for Truth

Siddhartha Gautama's decision to leave behind his life of privilege was not merely symbolic—it marked the beginning of a profound inward journey, one that would become the foundation of one of the world's great spiritual traditions.

In the silence of the forest and the company of ascetics, Siddhartha began his search.

He studied with the great spiritual teachers of his time, mastering deep meditative states known as **formless absorptions**. He learned yogic disciplines, practiced extreme austerities, and explored the outer edges of consciousness. He fasted until his body withered, pushed his physical form near death, and silenced his mind in the pursuit of liberation.

But for all his discipline, he found himself no closer to lasting freedom. The answers he sought—answers not just to *why we suffer*, but *how we awaken*—remained elusive.

In time, Siddhartha realized that neither indulgence nor extreme deprivation would lead to truth. Both were attachments in different forms—one to pleasure, the other to control. This insight would later crystallize into one of his core teachings: **the Middle Way**—a path of balance between denial and excess.

It was a revolutionary insight, not just philosophically, but practically. Siddhartha was discovering a spirituality rooted not in extremes, but in presence. Not in theory, but in direct experience.

According to tradition, after years of wandering and searching, Siddhartha sat beneath a fig tree—later known as the **Bodhi Tree**—in the town of Bodh Gaya, vowing not to rise until he had discovered the truth.

He did not ask the gods for deliverance. He did not return to familiar teachings. He simply observed—his breath, his thoughts, the nature of mind itself.

There, in stillness, he turned fully inward. And through deep contemplation, he began to see clearly: the impermanence of all things, the nature of craving and attachment, and the possibility of a mind unbound by fear or desire.

This was not intellectual realization. It was awakening—*bodhi*—a direct seeing into the way things are.

At the age of 35, Siddhartha became **the Buddha**, the Awakened One.

But unlike sages who retreat forever into silence, the Buddha chose to share his insight. He arose from his seat not to dominate, but to teach—not to convert, but to guide. His mission would not be to claim absolute truth, but to offer a **path** others could walk for themselves.

In the story of his renunciation and search, we find a model of spiritual integrity: a willingness to question every belief, to endure discomfort, and to remain open—even when the answers are not immediate.

The Buddha did not awaken by clinging to tradition or rejecting the world. He awakened by observing it, understanding it, and letting it be what it was.

Queries for Reflection

- What beliefs or identities must we leave behind to begin our own search for truth?
- Can spiritual insight arise not from doctrine, but from deep observation?
- Where in our lives do we oscillate between indulgence and deprivation—and what might the middle way look like?
- How do we recognize truth when it comes, and do we rise to share it?

5.2.3
Enlightenment Under the Bodhi Tree

It is said that when Siddhartha Gautama sat beneath the Bodhi Tree, he did not merely meditate—he resolved to confront the root of all suffering and illusion. He vowed to stay seated, unmoving, until he saw clearly. This moment—one of the most iconic in spiritual history—was not marked by divine fireworks or celestial pronouncements. It was marked by stillness, insight, and unwavering presence.

As the night deepened, **Māra**, the tempter, appeared—not as a devil in a Western sense, but as the personification of delusion and ego. Māra presented distractions: sensual desires, fears, self-doubt, and even lofty ideas. But Siddhartha remained unmoved. He saw these forces for what they were—passing phenomena, projections of the mind.

At one point, Māra challenged his right to seek liberation, demanding proof. In response, Siddhartha reached out his hand and touched the earth. This gesture, called the **Bhumisparsha mudra**, symbolizes calling the Earth as witness. Not to prove power, but to affirm presence. I am here. I see. I understand.

By dawn, the storm had passed—not outwardly, but inwardly.

In that moment of clarity, Siddhartha awakened. He saw the nature of existence as it truly is:

- That all conditioned things are **impermanent** (*anicca*)
- That clinging and craving give rise to **suffering** (*dukkha*)
- And that there is a way to become **free** (*nirvāṇa*)

This was not a revelation granted from above, but a realization uncovered from within. It did not rely on dogma or authority, but on direct perception. The Buddha did not claim to be a god, nor did he position himself as a savior. He had seen, and now he taught.

He had become awake—not to something new, but to what had always been present.

This enlightenment was not escape, but intimacy with reality. It was not an end, but a beginning. The Buddha now faced a choice: whether to keep this realization private or to share it with a world unlikely to understand.

According to the tradition, he hesitated. The truth he had seen was subtle, profound, and difficult to convey. But compassion moved him. He chose to teach—not as a master over others, but as a fellow traveler who had walked the path.

The Buddha would go on to teach for over 40 years, not asking for worship, but inviting inquiry. He offered no commandments, only observations. No demands for belief, only encouragement to test his words against one's own experience.

The enlightenment under the Bodhi Tree was not just a personal liberation—it became the seed of a living tradition. A tradition not of conversion, but of awakening. A tradition that would grow across time and culture, and still speak today to the deepest questions of the human heart.

Queries for Reflection

- What fears or doubts arise when we sit still with ourselves?
- How can we respond to the inner voices of distraction and resistance?
- What would it mean to "touch the earth" in our own lives?
- Is enlightenment a distant ideal, or a moment of clear seeing available here and now?

5.2.4
The Four Noble Truths and the Eightfold Path

After his awakening, the Buddha gave his first formal teaching in the Deer Park at Sarnath, near Varanasi. This discourse—known as the **Dhammacakkappavattana Sutta**, or "Turning the Wheel of Dharma"—introduced a framework that would become the cornerstone of Buddhist thought: the **Four Noble Truths**.

These truths are not dogmas to believe, but observations to explore—simple in language, yet profound in implication.

The First Noble Truth: Dukkha – Suffering Exists

Life contains suffering. This does not mean life is only suffering, but that dissatisfaction, unease, and impermanence are woven into existence. Pleasure fades. Loss comes. Even joy carries the shadow of its passing. Birth, aging, illness, death—none can be avoided.

The Buddha did not begin with despair, but with honesty. He invited us to look at the human condition not with fear, but with clarity.

The Second Noble Truth: Samudaya – The Cause of Suffering

Suffering arises from **craving**—from clinging to what we desire, resisting what we fear, and insisting on permanence in a world of change. We grasp at pleasure, identity, success, and certainty. We push away discomfort, loss, and insecurity. This grasping, rooted in ignorance, generates suffering.

But the cause is not external—it lies in our relationship to experience. And this insight brings possibility.

The Third Noble Truth: Nirodha – The Cessation of Suffering

If suffering has a cause, it can also have an end. When craving ceases, suffering ceases. This is **nirvāṇa**—not annihilation, but liberation. Not the absence of life, but the presence of peace. It is the unbinding of the mind from compulsive reactivity, and the return to spacious awareness.

This truth offers hope, not as a promise from above, but as a possibility within.

The Fourth Noble Truth: Magga – The Path to the Cessation

The Buddha did not stop at diagnosis. He offered a path—a way to walk toward freedom. This is the **Eightfold Path**, often grouped into three core areas: wisdom, ethical conduct, and mental discipline.

The Eightfold Path

1. **Right View** – Understanding the nature of reality and the Four Noble Truths
2. **Right Intention** – Cultivating intentions of renunciation, goodwill, and harmlessness
3. **Right Speech** – Speaking truthfully, kindly, and without harm
4. **Right Action** – Acting ethically, refraining from harm, theft, and misconduct
5. **Right Livelihood** – Earning a living in a way that does not exploit or injure
6. **Right Effort** – Directing energy toward wholesome states of mind
7. **Right Mindfulness** – Developing awareness of body, feeling, mind, and phenomena
8. **Right Concentration** – Cultivating deep meditative absorption

This path is not linear, nor a checklist. It is a **living framework**, to be practiced, refined, and adapted. It does not require belief in any deity or creed. It begins with observation, continues with discipline, and leads to transformation.

In the Eightfold Path, we see a holistic vision of human development—ethical, mental, and spiritual. It is not a set of commandments, but a way of life grounded in wisdom and compassion.

Queries for Reflection

- What does it mean to face suffering without denial or despair?
- How do our cravings—subtle or obvious—create discontent in our lives?
- Can we imagine a form of freedom that comes not from acquiring more, but needing less?
- What would it look like to walk the path—not as a philosophy, but as a practice?

5.2.5
The Middle Way and Impermanence

Before his enlightenment, Siddhartha had explored two extremes: the pleasures of royal life and the harsh asceticism of self-denial. Neither had brought him peace. From this lived experience arose a central insight that would shape his teaching: **the Middle Way**.

The Middle Way is not merely a compromise between indulgence and deprivation. It is a **principled path of balance**, of navigating life with discernment, mindfulness, and compassion. It honors the body without being enslaved to it. It cultivates discipline without falling into rigidity.

This path does not swing between attachment and aversion—it walks between them. It resists the illusion that happiness lies at the edges of experience and instead invites us into the heart of presence.

Alongside this vision of balance is another profound teaching: **impermanence**, or *anicca.*

Everything changes. Every sensation, every thought, every relationship, every structure—however solid it may seem—is in motion. The seasons shift, the body ages, emotions rise and fall. Even the self, when examined closely, reveals no fixed core—only a stream of conditions and perceptions flowing together moment by moment.

At first, impermanence can feel unsettling. We long for permanence, for things to stay as they are. We hold on to youth, to identity, to certainty. And when these inevitably change, we suffer.

But to see impermanence clearly is not to fall into despair. It is to awaken from illusion. It is to recognize that life is not a static possession, but a dynamic unfolding. The more we cling, the more we suffer. The more we let go, the more we can love freely, act wisely, and rest in the truth of what is.

The Middle Way and impermanence are not abstract doctrines—they are lenses through which to see life. They guide us to respond, not react; to adapt, not grasp; to meet the moment with open hands.

In a world obsessed with extremes—of wealth and poverty, belief and rejection, pleasure and pain—the Middle Way calls us back to balance.

In a world that idolizes permanence—of fame, security, and form—impermanence invites us into deeper humility and gentleness.

Together, these teachings offer not an escape from life, but a new way to live within it.

Queries for Reflection

- Where in our lives do we swing between extremes, and what might balance look like?
- How do we respond when what we love begins to change—or ends?
- Can impermanence be seen not as a threat, but as a teacher?
- What does it mean to walk the Middle Way in a time of polarization and grasping?

5.2.6
Buddhism Across Time and Culture

After his enlightenment, the Buddha traveled and taught for more than 40 years, gathering disciples and planting seeds of wisdom across northern India. Yet he left no temple, no scripture penned in his own hand, no central institution. What he left was a path—a way of seeing, living, and awakening—that would travel far beyond its original home.

Within a few centuries, the teachings of the Buddha had begun to spread across Asia, taking root in diverse cultural soils. What emerged was not a monolithic religion, but a family of traditions—each expressing the Dharma in its own language, symbols, and forms.

Theravāda: The Way of the Elders

The **Theravāda** tradition, prevalent in countries like Sri Lanka, Thailand, and Myanmar, emphasizes the earliest recorded teachings of the Buddha, preserved in the *Pāli Canon.* It focuses on individual liberation through meditation, ethical living, and insight into impermanence and non-self.

The ideal figure here is the **arahant**—one who has overcome ignorance and attained freedom. Theravāda Buddhism is often known for its clarity, discipline, and deep commitment to preserving the original path.

Mahāyāna: The Great Vehicle

Emerging a few centuries later, **Mahāyāna Buddhism** spread across China, Korea, Vietnam, and later Japan. It introduced a more expansive vision of the path—not just personal liberation, but universal awakening.

The Mahāyāna ideal is the **bodhisattva**—a being who, though capable of entering nirvāṇa, chooses to remain in the world out of compassion for all beings. This tradition gave rise to profound philosophical developments (like emptiness and dependent origination), as well as devotional practices centered around celestial Buddhas and bodhisattvas such as Avalokiteśvara and Amitābha.

Mahāyāna Buddhism is diverse and poetic, often emphasizing the inseparability of wisdom and compassion.

Vajrayāna: The Diamond Vehicle

Vajrayāna, or Tantric Buddhism, developed within Mahāyāna and found its most vivid expression in Tibet, Bhutan, and parts of Mongolia. It incorporates rituals, mantras, visualization practices, and symbolic imagery aimed at transforming consciousness swiftly and directly.

This path embraces the full range of human experience as fuel for awakening. It sees even the most challenging emotions as potential gateways to realization when skillfully understood.

Figures like **Padmasambhava** and the line of Tibetan **lamas** represent this living tradition, which combines scholarly depth, mystical symbolism, and deep devotion.

Zen and Beyond

In Japan, **Zen Buddhism** emerged as a stripped-down, experiential form of Mahāyāna. Zen emphasizes direct insight through meditation (*zazen*), simplicity, and paradoxical teachings known as **koans**. "What is the sound of one hand clapping?" Such questions are meant not to be answered, but to dissolve conceptual thinking and awaken intuitive knowing.

Elsewhere, Buddhism merged with Indigenous traditions, adapted to new cultures, and took on local languages and aesthetics. In the West, it has often been interpreted through a psychological lens—emphasizing mindfulness, compassion, and presence over metaphysics or ritual.

A Living Tradition

What makes Buddhism unique is not its uniformity, but its adaptability. It has moved across continents and millennia not by conquest, but by resonance.

Though teachings and practices differ, the essence remains: the possibility of awakening, the path of compassion, and the liberation from suffering through understanding.

In today's global world, Buddhism continues to evolve—shaped by science, dialogue with other faiths, and the inner journeys of millions. It remains not a relic of the past, but a dynamic and living inquiry into the nature of mind and reality.

Queries for Reflection

- What does it mean for a spiritual path to evolve across cultures without losing its core?
- Can different expressions of the same wisdom coexist without conflict?
- How do we hold both reverence for tradition and openness to change?
- In our own lives, how do we translate ancient insight into modern practice?

5.2.7
The Buddha in Interfaith and Contemporary Thought

Though the Buddha lived over 2,500 years ago, his presence has not been confined to temples or textbooks. In modern times, his image, teachings, and example have transcended religious boundaries—resonating with seekers, scholars, and spiritual traditions around the world.

In an age marked by religious pluralism and global dialogue, the Buddha is often seen not only as the founder of a religion, but as a **universal figure of wisdom**. His insights into the nature of mind, suffering, and compassion are increasingly viewed as relevant across all belief systems—or even outside them entirely.

The Buddha and the Modern Mind

The psychological clarity of the Buddha's teachings has drawn the interest of scientists, philosophers, and psychologists. Concepts such as **impermanence**, **non-attachment**, and **mindfulness** align closely with contemporary understandings of neuroplasticity, cognitive behavioral therapy, and emotional regulation.

The **mindfulness movement**, though often secularized, has its roots in the Buddha's emphasis on moment-to-moment awareness (*sati*). Practices like meditation, once seen as esoteric, are now widespread in schools, hospitals, and workplaces—not as religious exercises, but as tools for well-being.

Many therapists and researchers draw upon Buddhist insights to explore how identity is constructed, how thought patterns perpetuate suffering, and how awareness can lead to personal transformation. The

Buddhist path is increasingly appreciated not as a belief system, but as a method of inquiry and self-understanding.

The Buddha in Other Religions

In **Hinduism**, the Buddha is often recognized as an avatar of Vishnu—one who came to restore dharma in a time of spiritual confusion. Though the doctrinal distinctions between Buddhism and Hinduism remain, there is also mutual reverence.

In **Christianity**, some have drawn parallels between Jesus and the Buddha—not in doctrine, but in spirit. Both taught love, compassion, renunciation of worldly attachments, and the value of the inner life. Christian contemplatives, including Thomas Merton and Henri Nouwen, engaged deeply with Buddhist teachings, finding shared ground in silence, humility, and transformation.

In **Islam**, the Buddha is not explicitly mentioned in the Qur'an, but some Islamic scholars and mystics have acknowledged him as a wise teacher or even a prophet-like figure. In the mystical tradition of **Sufism**, themes of ego dissolution, divine love, and inward journeying mirror aspects of the Buddhist path.

In **Judaism**, particularly among modern Jewish mystics and Renewal movements, the Buddha's teachings have sparked rich dialogue about mindfulness, suffering, and the cultivation of presence. Many Jewish practitioners now explore **dual belonging**—embracing Buddhist practice without abandoning Jewish identity.

A Figure Beyond Borders

The Buddha's image—serene, seated, eyes half-closed in deep presence—has become iconic. But what he represents extends beyond culture or creed. He symbolizes the possibility that awakening is not granted from above, but discovered within.

His life reminds us that truth does not belong to one tradition. It can emerge wherever there is sincerity, effort, and insight. His teachings remind us that we are not separate from the patterns we seek to understand. And his silence reminds us that the deepest truths are sometimes beyond words.

In an era of fragmentation and overload, the clarity and calm of the Buddha's path speak to something timeless: that peace is not a distant achievement, but a way of being available now.

Queries for Refl ection

- How has the Buddha's legacy been reinterpreted in the modern world—for better or worse?
- Can insights from one tradition enrich others without diluting their depth?
- Is it possible to approach the Buddha not only as a religious founder, but as a mirror of the awakened potential in all?
- How do we engage ancient teachings with fresh eyes and open hearts?

5.2.8
Awakening the Inner Buddha

For all the reverence shown to Siddhartha Gautama, the heart of his teaching was not self-worship, but self-transcendence. The Buddha did not ask to be followed blindly—he asked that his path be tested, practiced, and realized. He did not say, "Worship me," but rather, "Be a light unto yourselves."

This is one of the most radical messages in all spiritual history: that **awakening is not exclusive**. It is not the possession of a priesthood or a reward for belief. It is a **potential already within**—inherent in the human mind, waiting to be uncovered.

In Mahāyāna Buddhism, this is expressed through the teaching of **Buddha-nature**—the idea that beneath our confusion, fear, and desire lies an unshakable clarity and compassion. Like clouds obscuring the sun, our suffering is not who we are. It is what we've mistaken ourselves to be.

To awaken the inner Buddha is not to become something new, but to **remember** what has always been present. It begins not with striving, but with stillness. Not with perfection, but with honest attention.

The path may involve meditation, ethical living, or philosophical study—but its essence is simple: to be here, now, with open eyes and a soft heart.

In this sense, the Buddha's story is not just a biography—it is a map.

- The palace is our comfort zone.
- The four sights are our moments of awakening.
- The forest is our period of seeking.
- The Bodhi Tree is our moment of surrender.
- And the awakening is the deep seeing into what has always been true.

This pattern does not belong to ancient India—it belongs to every soul who begins to wonder: is there more than this? And if there is, how do I find it?

To walk this path is not to become Buddhist, but to become **awake**—to live without clinging, to love without condition, and to meet each moment with mindful presence.

In a time when speed and stimulation dominate, the Buddha's example offers an alternative: silence, simplicity, clarity. His legacy is not bound by temples or labels—it lives wherever a person pauses, breathes, and begins to truly see.

Final Queries for 5.2

- What does it mean to awaken—not someday, but now?
- How can we relate to the Buddha's story as a mirror of our own journey?
- Is enlightenment reserved for the rare few, or available in ordinary moments of insight?
- What might change if we lived as if the Buddha-nature within us was already real?

3.
KRISHNA AND RAMA

Divine Incarnations of Dharma and Love

-ZEPHYRO SOLON

CONTENTS

5.3.1
Avatars and the Cosmic Framework of Dharma

In the vast and ancient spiritual landscape of Hinduism, few ideas are as central—and as layered—as the concept of the **avatar**. Unlike figures who arise solely from historical events, avatars are believed to descend from the divine, taking form in our world not because they must, but because they choose to. Their purpose: to restore balance, protect the righteous, and reawaken dharma.

The word **avatar** comes from the Sanskrit *ava* (down) and *tara* (to cross)—literally, "one who crosses down." In this tradition, God is not separate from the world, nor indifferent to its struggles. When darkness rises and imbalance prevails, the divine descends—not as abstract force, but as embodied presence.

The **Bhagavata Purana** and other sacred texts speak of many such incarnations, each suited to the needs of its age. Some are cosmic and symbolic—like the fish that saves humanity from a flood, or the boar that raises the Earth from the depths. Others are fully human in appearance, yet carry the fragrance of divinity.

Among these, **Rama** and **Krishna** stand as two of the most beloved and enduring. Both are considered full incarnations (*purna avatars*) of Vishnu, the aspect of the divine responsible for preservation and cosmic order. Yet they manifest very differently.

- **Rama** is the embodiment of honor, righteousness, and kingship. He is the dutiful son, the faithful husband, the noble warrior. His life exemplifies the ideal of *dharma*—not merely as law, but as cosmic harmony reflected through action.
- **Krishna**, on the other hand, is the embodiment of joy, play, and divine love. He is the mischievous child, the playful flute-player, the wise

charioteer. He transcends conventions and reveals the sacred through intimacy, beauty, and paradox.

Together, they represent two faces of the divine:

- **Rama**, the path of responsibility and structure
- **Krishna**, the path of surrender and love

In Hindu cosmology, time moves in great cycles called **yugas**. Each yuga brings its own challenges and spiritual climate. The avatar appears not to create a new truth, but to restore what has been forgotten. In this view, divine intervention is not a disruption of nature, but its renewal.

The stories of Krishna and Rama are not only ancient epics; they are living scriptures. They are recited in homes, temples, and festivals, not as distant myths, but as mirrors of the human journey.

They remind us that the divine does not always arrive in thunder or flame. Sometimes, it walks beside us, speaks in riddles, weeps in exile, or plays a flute beneath the moon.

To study Krishna and Rama is not only to learn about gods of India—it is to explore how divinity moves through time, form, and relationship. It is to ask: What would it mean if the sacred chose to walk among us? And what would it mean if the purpose of that presence was not to dominate, but to restore balance?

Queries for Reflection

- What does the idea of an avatar suggest about the relationship between the divine and the world?
- Can divinity take human form without losing its essence?
- Do different ages require different kinds of spiritual guidance?
- What part of ourselves is drawn to Rama's discipline—and what part to Krishna's joy?

5.3.2

Rama: The Ideal King and the Path of Duty

To understand Rama is to enter the realm of **dharma**—not simply as moral conduct, but as cosmic alignment. In Hindu tradition, Rama is not only a heroic figure; he is the living embodiment of righteousness (*Maryada Purushottama*)—the "perfect man" who reveals what it means to live with integrity, even when it costs everything.

The story of Rama is told in the **Ramayana**, one of India's great epics, traditionally attributed to the sage Valmiki. More than a tale of war or romance, the Ramayana is a spiritual map, illustrating how the divine moves through human relationships, trials, and choices.

Rama is born as the prince of Ayodhya, the eldest son of King Dasharatha. From a young age, he displays nobility, restraint, and deep inner calm. He wields weapons, yes—but never out of anger or ambition. His strength is matched by humility. His power is rooted in principle.

When his father prepares to crown him king, a twist of fate changes everything. Bound by a promise to one of his queens, the king is compelled to exile Rama for fourteen years and install another son in his place. Rama does not resist. He does not argue. He accepts the exile with quiet grace—not because he is weak, but because **duty transcends desire**.

His wife, **Sita**, and his loyal brother, **Lakshmana**, choose to accompany him. Together they live in the forest, facing hardship, demons, and moral tests. When Sita is abducted by the demon-king Ravana, Rama's path turns toward battle. Yet even in war, he remains aligned with dharma—fighting not for revenge, but for justice and restoration.

When Ravana is finally defeated and Sita is rescued, a deeper trial begins: the trial of public scrutiny, of reputation, of trust. Rama, though heartbroken, asks Sita to prove her purity—a choice that continues to be debated across time. For some, it shows the cost of ideal kingship in a flawed world. For others, it reveals the tension between personal love and public responsibility.

Rama's story is not simple. It is filled with **noble pain**—the kind that arises when one chooses what is right over what is easy. He renounces power, love, and even his own comfort to uphold a larger order.

In Rama, we see a model of **ethical leadership**—not power for its own sake, but service to something higher. He does not seek adoration. He seeks alignment. And in doing so, he reveals a truth that echoes across traditions: that freedom is not always the absence of duty, but its fulfillment with grace.

For many, Rama is not just a historical or mythological figure. He is the quiet voice within that asks us to choose honor when the world pulls toward compromise. He is the presence that reminds us that strength can be tender, and that principle matters—even when no one is watching.

Queries for Reflection

- What does it mean to live in alignment with dharma in our time?
- Can we admire Rama's choices even if we struggle with their consequences?
- How do we balance personal loyalty and public responsibility?
- What part of ourselves longs to act with such clarity, even when it hurts?

5.3.3
The Ramayana: Exile, Trial and Return

The story of Rama unfolds in the **Ramayana**, a sacred epic that is at once poetic scripture, cultural foundation, and spiritual parable. Across centuries and continents, it has been recited in temples, sung in marketplaces, and adapted into countless languages and art forms. Yet beneath its grandeur lies a profoundly human story—of exile, love, conflict, and return.

The narrative begins in the golden city of **Ayodhya**, where Rama, the beloved prince, is poised to inherit the throne. But through a complex web of palace politics and a vow made long ago by his father, Rama is sent into **exile for fourteen years**. His younger brother **Bharata**, though offered the throne, refuses to rule in Rama's place—leaving Rama's sandals on the throne as a symbol of rightful kingship.

This exile is not just physical—it is symbolic. It marks a descent into the wilderness, where the outer structures of status and comfort fall away. Here, Rama, Sita, and Lakshmana live close to the earth, encountering sages, demons, and the elemental forces of nature. It is a time of testing, transformation, and hidden revelation.

The heart of the Ramayana turns when **Sita is abducted** by the demon-king **Ravana**, ruler of the distant island of Lanka. Her kidnapping sets in motion one of the great rescue missions of world literature. Rama allies with the vanaras—monkey-like beings led by the mighty **Hanuman**, whose devotion becomes legendary.

Hanuman's leap across the ocean to find Sita, and his burning of Lanka to deliver Rama's message, is one of the most beloved portions of the epic. It represents not just courage, but unwavering devotion—*bhakti*—to a higher cause.

The **battle between Rama and Ravana** is not just a war of armies, but a clash of principles: dharma versus ego, justice versus arrogance, restraint versus lust for power. Ravana is no simple villain—he is learned, cultured, and proud. But his refusal to release Sita reveals the downfall of even the most brilliant mind when untempered by humility.

When Rama finally defeats Ravana and rescues Sita, the reunion is bittersweet. Questions of honor, purity, and public perception arise. Though Sita has remained faithful, Rama asks her to undergo a **trial by fire** to prove her chastity. She enters the flames—and emerges unharmed, a testimony to her purity. And yet, in some versions of the story, this pain is not the end.

Later in their reign, when whispers of doubt spread among the people, Rama makes the agonizing choice to **banish Sita** again—this time, pregnant and alone. She finds refuge in a hermitage, where she gives birth to twin sons. Years later, in a dramatic reunion, the family is brought together again—but Sita, weary of the world's judgment, chooses to return to the Earth, her mother.

The **Ramayana ends with Rama's return** to the heavens, his mission complete. He is not only a king, but an avatar—an embodiment of divine order, whose life story remains a map for ethical struggle and inner strength.

But the Ramayana is more than the story of one man. It is a mirror of life's journey:

- The **exile** we face when the world no longer fits us.
- The **trials** that ask us to act with integrity under pressure.
- The **devotion** that sustains us through darkness.
- The **return** that may not restore what was lost, but opens a new wholeness.

Rama's story is not easy. It demands thought. It raises questions. It invites us to reflect not only on heroism, but on cost, choice, and consequence.

Queries for Reflection

- What parts of our own journey mirror exile, trial, or return?
- Can dharma sometimes conflict with desire—and if so, which do we follow?
- How do we reconcile compassion with public duty, or love with principle?
- Is the Ramayana a distant legend—or a mythic telling of universal human experience?

5.3.4
Krishna: Divine Play and the Embodiment of Love

If Rama represents divine order, Krishna reveals divine **play**. Where Rama walks the straight line of duty, Krishna dances through paradox. Where Rama is the noble king, Krishna is the cosmic child, the mischievous lover, the strategist in war, and the teacher of eternal wisdom. He embodies **leela**—the divine playfulness through which the universe unfolds.

Krishna's story is told across many texts—the *Mahabharata*, *Bhagavata Purana*, *Harivamsa*, and countless regional songs and folk traditions. To reduce him to a single role is impossible. He is, at once, **fully human and fully divine**—a cowherd and a kingmaker, a flute player and a philosopher, a trickster and a guide.

Born in the shadow of danger, Krishna is said to have come into the world to defeat the tyrant Kamsa. His birth story is miraculous—delivered in a prison cell, carried across a flooded river to safety. Even as a child, his presence disrupts evil and awakens joy.

As a youth in **Vrindavan**, Krishna enchants all with his beauty, music, and light-hearted mischief. He steals butter, plays tricks, and dances with the **gopis**—the milkmaids who are drawn to him not by command, but by the irresistible pull of divine love. Among them, **Radha** stands out—not as a mere romantic figure, but as a symbol of the soul's longing for union with the Infinite.

Krishna's flute, in this context, becomes more than a musical instrument—it becomes the **call of the divine**, beckoning us away from worldly distraction and toward ecstatic surrender.

Later, Krishna becomes a central figure in the **Mahabharata**, guiding the Pandava brothers in their struggle for justice. As a political advisor, he helps navigate the complexities of dharma in a broken world. But it is on the eve of battle—when the warrior Arjuna hesitates—that Krishna reveals his deepest role: that of **teacher and revealer of truth**.

In the **Bhagavad Gita**, Krishna speaks not as a partisan, but as the voice of eternal wisdom. He explains the nature of the soul, the illusion of death, the importance of action without attachment, and the paths of knowledge (*jnana*), devotion (*bhakti*), and selfless service (*karma*). When Arjuna sees Krishna's true form—radiant, cosmic, terrifying and beautiful—he beholds not a man, but the entire universe in motion.

Krishna's uniqueness lies in his **embodiment of opposites**:

- He is playful yet profound
- Detached yet deeply loving
- Involved in the world, yet never bound by it

He shows that divinity does not only reside in the ascetic's cave or the king's palace—but in the dance of life itself. He challenges the notion that the sacred must be solemn. In Krishna, we find a spirituality that laughs, sings, weeps, and embraces the mess of being human.

To know Krishna is not only to read about him, but to be drawn into relationship—through love, through devotion, through surrender. His stories invite us to go beyond the mind and enter the heart, where truth is not explained, but felt.

Queries for Reflection

- Can spiritual wisdom coexist with joy, beauty, and play?
- What does Krishna's story teach us about love that is not possessive, but freeing?
- How do we respond when the divine appears not in seriousness, but in song?
- Is it possible that play, too, can be sacred?

5.3.5
The Bhagavad Gita: Action, Surrender, and Inner War

Among all the sacred texts of Hinduism, the **Bhagavad Gita** holds a special place—as both battlefield dialogue and timeless spiritual teaching. Nestled within the vast epic of the *Mahabharata*, the Gita unfolds at a moment of profound crisis. Yet its setting—on the verge of war—is less about politics and more about the **inner conflict** that lives in every human soul.

The warrior **Arjuna**, noble and skilled, stands paralyzed between two armies. On one side are his enemies; on the other, his kin. As he surveys the battlefield, he is overcome—not with fear of death, but with sorrow over duty. How can he fight those he loves? How can war be righteous?

His **charioteer**, Krishna, is no ordinary guide. In a moment of hesitation, Krishna reveals not only wisdom, but his divine nature. The ensuing dialogue becomes the **Bhagavad Gita**—a scripture of only 700 verses that speaks to the depths of human struggle.

Karma Yoga – The Path of Action

Krishna begins by urging Arjuna to **act**—not from anger, greed, or ambition, but from **selfless duty**. He teaches that one must perform their dharma, even when the heart is torn. What matters is not the outcome, but the **intention** and **detachment** behind the act.

"You have the right to action, but not to the fruits of action."

This is the essence of **karma yoga**—acting without clinging to success or failure, praise or blame. It is not inaction, but *inner freedom in the midst of action.*

Bhakti Yoga – The Path of Devotion

As the dialogue deepens, Krishna invites Arjuna to go beyond duty and enter the path of **love**. This is **bhakti yoga**—devotion to the Divine not as an abstract force, but as a living presence. To see Krishna not only as a god, but as the indwelling guide—the inner Self beyond ego and mind.

"Offer me your heart, your devotion, your sacrifices. I will accept you and lift you from all sin."

In bhakti, surrender is not weakness, but strength—the release of control into the arms of something greater.

Jnana Yoga – The Path of Knowledge

Krishna also offers the wisdom of **jnana yoga**, the path of discernment. He teaches that the soul (*atman*) is eternal, unborn, and undying. What we call death is simply the shedding of a form. What we fear is illusion.

"As a person casts off worn-out clothes and puts on new ones, so the soul discards the body and takes on another."

In this teaching, we are reminded that the Self is not the body, the role, or even the story. It is the silent witness behind all experience.

A War Within

Though the Gita is spoken on the eve of battle, it is not a call to violence—it is a call to **clarity**. The true war is not between rival clans, but within the human heart: between fear and courage, attachment and freedom, self-interest and surrender.

Arjuna's hesitation is universal. We all stand, at times, at the edge of inner conflict—unsure how to act, unsure who we are. The Gita offers not fixed rules, but a framework for living with depth, devotion, and discernment.

Its vision is vast. One can worship God in form or formlessness, through silence or service, meditation or music. The Divine meets each soul where it is.

In Krishna's final words, he invites Arjuna—and us—not into blind obedience, but into inner alignment:

"Reflect on this fully, then do as you choose."

Queries for Reflection

- What inner battle are you facing, and what wisdom might guide your action?
- Can we act with full engagement, yet without attachment to results?
- What does it mean to surrender—not to fate, but to a deeper flow of life?
- How might love, clarity, and duty converge into one path?

5.3.6
Devotion (Bhakti) and the Personal Face of God

In the vast forest of Hindu spiritual paths, **bhakti**—the path of devotion—shines with a unique warmth. While other approaches may emphasize discipline, knowledge, or renunciation, bhakti is rooted in relationship. It is not about escaping the world, but about loving the Divine *within* it, and *through* it.

At the center of bhakti lies one core truth: **God is not distant**. God is not abstract. God has a face, a voice, a heart—and that heart is turned toward us.

This understanding finds its fullest expression in the lives and legacies of Krishna and Rama.

The Devotion to Rama

In Rama, the devotee sees the **ideal king**, the moral compass, the protector of righteousness. His devotees, like **Hanuman**, embody unwavering loyalty and selfless service. Hanuman does not merely admire Rama—he lives for him. He fights for him. He prays, not to gain heaven, but simply to be near him.

Hanuman's love is pure bhakti: no bargaining, no fear—only presence.

Rama's devotees sing hymns (*bhajans*), chant his name, and meditate on his image—not out of superstition, but to align their hearts with his clarity and strength. In their devotion, they find not just reverence, but inner resilience.

The Devotion to Krishna

Krishna's bhakti takes on a more intimate and playful form. He invites the soul not into servitude, but into **relationship**—as a friend, a beloved, a child, even a cosmic companion.

His most devoted followers, like **Radha**, do not seek liberation—they seek union. Radha's love for Krishna is not metaphor alone; it represents the **soul's yearning** for the Infinite. Her longing, her tears, her surrender—they become the scripture of the heart.

In bhakti, even pain is sacred. Separation from the Beloved intensifies the fire of love. This is not sentimentality—it is transformation through tenderness.

Bhakti as a Universal Path

Bhakti is not limited by language or lineage. It has flourished in temples and fields, in Sanskrit verses and folk songs. Saints like **Mirabai**, **Tulsidas**, **Surdas**, and **Kabir** have poured their devotion into poems that still stir the soul.

Some devotees speak to God as father or mother, others as child or teacher. Some see God in the form of Krishna, others in Shiva, Durga, or the formless Absolute. The essence remains the same: a **heart turned toward the Divine**, like a flower to the sun.

Bhakti dissolves hierarchy. It does not demand scholarship or asceticism. Even the illiterate, even the poor, even the broken-hearted can walk this path. Sometimes, especially the broken-hearted.

In Krishna and Rama, the personal face of God becomes tangible. One can love them. One can cry to them. One can dance with them. The Divine becomes not just the goal, but the companion on the journey.

Queries for Reflection

- What does it mean to approach the Divine not with fear, but with love?
- Can devotion be a practice, not of dogma, but of relationship?
- What image or presence of the sacred draws your heart forward?
- Is it possible that longing itself is a doorway to the Divine?

5.3.7
Krishna and Rama in Comparative and Contemporary Perspective

Krishna and Rama are not merely figures of the past. They continue to live in story, ritual, art, and memory—not just as icons, but as **archetypes** of the spiritual journey. Their significance extends beyond Hinduism, reaching into philosophy, psychology, and interfaith dialogue. Each reflects a different facet of divinity—and, by extension, of human possibility.

Two Paths, One Essence

Rama and **Krishna** both serve as avatars of Vishnu, yet their expressions differ.

- **Rama** represents **structure, dharma, and moral clarity**. His life is a blueprint for duty, restraint, and ethical sovereignty. He asks us to act rightly, even when it hurts.
- **Krishna** represents **freedom, love, and divine spontaneity**. His life dissolves rigid boundaries, revealing the sacred in joy, beauty, and paradox. He asks us to let go and trust.

In many ways, Rama shows us **how to live**, while Krishna shows us **how to live free**. One calls us to responsibility; the other to surrender. One draws us through principle, the other through play.

Yet both lead to the same source. Both are rooted in **presence**, in the unfolding of divine intelligence through form.

Psychological and Archetypal Views

From a modern psychological lens, these figures can be seen as **expressions of the inner self**:

- Rama may reflect the **conscious self**—the part of us that strives for balance, responsibility, and order.
- Krishna may reflect the **transcendent self**—the part that knows joy beyond circumstance and sees through the illusions of the world.

Together, they reflect the dual need within us: for rootedness and release, for purpose and play, for righteousness and grace.

In Jungian terms, they are **archetypes** of the *hero* and the *divine child*, the *sovereign* and the *trickster*, unified not in opposition but in complement.

In Interfaith and Global Discourse

Outside of Hindu tradition, Rama and Krishna have been honored by poets, saints, and philosophers from many backgrounds. Gandhi, for instance, revered **Rama's name** as a source of strength and peace. His final words—"*Hey Ram*"—were not political, but devotional.

Krishna's teachings in the *Bhagavad Gita* have inspired Christian mystics, Islamic Sufis, Jewish scholars, and Western philosophers. Aldous Huxley, Carl Jung, and Joseph Campbell all drew from Krishna's dialogue with Arjuna as a universal model for human transformation.

Even within contemporary yoga, mindfulness, and consciousness studies, the stories of Krishna and Rama offer rich ground. Whether seen as literal beings, symbolic myths, or inner guides, they continue to shape how millions perceive the sacred.

A Living Presence

To many devotees today, Krishna and Rama are not symbolic—they are **real**. They are prayed to, sung to, dreamed of. They appear in visions, in stories told to children, in mantras whispered in the dark.

Their relevance is not just spiritual but ethical. In an age of confusion and crisis, the call of dharma—echoing from Rama—and the invitation to joyful surrender—offered by Krishna—remain deeply needed.

Their dual presence reminds us that the Divine is not bound by a single form or personality. Sometimes it comes in silence and discipline. Sometimes in laughter and music. But it always comes with love.

Queries for Reflection

- Which qualities of Rama or Krishna most resonate with your current path?
- Can duty and delight coexist in a full spiritual life?
- What modern figures or experiences echo these ancient avatars?
- How do these stories guide us—not only in worship, but in living well?

5.3.8
Living the Avatar Within

The stories of Krishna and Rama are not just to be admired—they are to be **embodied**. Not by imitation, but by **realization**. For while these avatars are said to descend from the divine, their deeper purpose is to **awaken the divine potential in us**.

This is the secret beneath the mythology: that every soul carries a spark of the same consciousness that moved through Krishna's flute and Rama's bow. That within each of us lies a capacity for wisdom, courage, compassion, and clarity.

In times of chaos, we can walk like Rama—with dignity, integrity, and restraint.

In times of doubt or fear, we can act like Arjuna, asking questions until our path becomes clear. And in times of beauty, intimacy, and joy, we can remember Krishna's smile—the reminder that **God dances, too**.

To **live the avatar within** does not mean perfection. It means alignment. It means living not as a slave to fear or desire, but as a conscious participant in something greater—call it dharma, divinity, or the flow of life itself.

This is not a burden. It is an invitation.

We may never lift a bow like Rama or deliver a discourse like Krishna, but we can reflect their essence in simple acts:

- Choosing truth when it costs us
- Loving without condition
- Serving without ego
- Acting without attachment
- Dancing through life with reverence and delight

The avatars descended into human form not to prove superiority, but to walk with us—through the forest, through the battlefield, through the village square. They remind us that the sacred does not hover above; it pulses within.

And if they descended into form to restore dharma, perhaps our role is to **raise it**—to lift our daily lives into alignment with love, with courage, with clarity.

Not because we must, but because we can.

Final Queries for 5.3

- What would it mean to live as if the divine were already within you?
- Which qualities of the avatar—Rama's steadiness or Krishna's joy—do you most need now?
- Can your ordinary life become a sacred path of devotion, action, and play?
- What if the stories of the avatars were not just about them—but about you?

4.

MOSES AND THE PROPHETS

Revealing the Will of God Across Generations

-ZEPHYRO SOLON

CONTENTS

5.4.1
The World of the Covenant: Israel's Story Begins

Long before the thundering voice on Sinai or the parting of seas, the story of Moses and the prophets begins with something quieter—a covenant. A whispered promise between the Infinite and a wandering people. A thread woven through time, binding heaven and earth in a relationship of purpose, struggle, and sacred destiny.

This is the story of **Israel**, a name that means "one who wrestles with God." It is not a tale of perfection, but of perseverance. Not of conquest alone, but of conscience.

The Hebrew Bible begins not with Moses, but with Abraham—a man who hears a call to leave his homeland and trust an unseen path. His descendants, through Isaac and Jacob, become a family, then a people. They endure famine, migrate to Egypt, and, over generations, become **enslaved** under a regime that forgets their origins.

It is into this world—one of oppression, memory, and longing—that Moses is born.

Yet even before Moses, there is a sense of divine intention unfolding. The God of the Hebrew Bible is not distant or indifferent, but **involved**—calling, guiding, testing, and remembering.

This is not the abstract principle of Eastern metaphysics or the formless stillness of the Tao. It is a **relational God**—one who speaks, commands, forgives, and sometimes contends.

Central to this worldview is the idea of **covenant** (*berith*)—a sacred agreement between God and humanity. With Noah, it is a promise never again to destroy the earth. With Abraham, a promise of descendants, land, and blessing. With the Israelites at Sinai, a binding law that will define a people set apart.

The Hebrew scriptures are not only a record of events—they are a dialogue. A conversation between the Divine and a community that is asked not just to believe, but to **remember**, to **obey**, and above all, to **listen**.

"Hear, O Israel: the Lord is our God, the Lord is One." (*Shema Yisrael* —Deuteronomy 6:4)

To be part of this covenant is to be part of a story. A story of liberation and law, exile and return, sin and renewal. It is a story not only of individuals, but of a people shaped by **calling**.

In this tradition, holiness is not abstract. It is lived—through justice, Sabbath, dietary law, ethical speech, and community care. The sacred is not far away—it is woven into daily life.

Before Moses appears, this is the world he inherits: a people enslaved, a God who remembers, and a promise still unfolding.

And then the bush begins to burn.

Queries for Reflection

- What does it mean to be part of a covenant—chosen not for superiority, but for responsibility?
- How does the idea of a relational God differ from more abstract visions of the divine?
- Can a people's history itself be a sacred text?
- What happens when the sacred is found not only in ritual, but in justice and memory?

5.4.2
Moses and the Burning Bush: A Call to Liberation

Moses' story begins in tension: born into a Hebrew family, raised in the Egyptian palace, caught between two worlds. As an infant, he is placed in a basket on the Nile to escape Pharaoh's decree to kill Hebrew male children. He is drawn from the water—his name, *Moshe*, means "to draw out"—and grows up as both outsider and insider, Hebrew and Egyptian, child of suffering and child of privilege.

This dual identity sets the stage for everything that follows.

As a young man, Moses witnesses an Egyptian beating a Hebrew slave. Overcome by rage, he strikes the oppressor and kills him. It is a moment of justice—but also of rupture. Forced to flee into exile, Moses ends up in the land of Midian, where he marries, becomes a shepherd, and begins a quieter life far from empire and memory.

And then, one day, in the stillness of the desert, **he sees a bush that burns but is not consumed**.

He turns aside to look.

This simple act—of noticing, of being drawn by wonder—becomes the hinge of destiny. In that moment, the ordinary is pierced by the sacred. From the flames comes a voice: **"Moses, Moses."**

He answers, "Here I am."

The voice identifies itself not as an abstract force, but as the God of his ancestors—of Abraham, Isaac, and Jacob. A God who has **seen the suffering** of the Hebrews and has come to deliver them. A God who calls Moses to lead.

Moses resists. He feels inadequate, unworthy. "Who am I," he asks, "that I should go to Pharaoh?" Even when promised divine support, he hesitates. "What is your name?" he asks.

The reply is enigmatic and vast:

"Ehyeh asher ehyeh"
I Am That I Am.
Or, *I Will Be What I Will Be.*

This is not a name in the conventional sense—it is a **revelation of being itself**. A God who is not fixed, not contained. A Presence that is dynamic, unfolding, and deeply personal.

Still unsure, Moses is given signs—his staff turns to a serpent, his hand is made leprous and healed. But he still pleads for another to go in his place. Finally, God allows his brother Aaron to accompany him.

This encounter—mystical, terrifying, intimate—is a turning point in spiritual history. It is not only the call of a prophet, but the revelation of a new kind of relationship between the Divine and humanity: one rooted not just in worship, but in **liberation**.

God is not only the Creator of the cosmos but the Deliverer of the oppressed. A God who hears cries, who sees injustice, who acts through chosen messengers to challenge the powers of the world.

And Moses, reluctant yet chosen, becomes the vessel of that call.

The burning bush reminds us that the sacred does not always arrive with thunder. Sometimes it flickers in the corner of our awareness—waiting for us to turn aside and look.

Queries for Reflection

- Have you ever encountered a "burning bush"—a moment of quiet revelation that changed everything?
- What does it mean that Moses, even in doubt, was still called?
- How does this story redefine the Divine—not only as transcendent, but as responsive and engaged?
- In a world of injustice, what might it mean to be a liberator—not through power, but through purpose?

5.4.3
The Exodus: Journey Through Wilderness and Wonder

The story of the **Exodus** is one of the most defining spiritual narratives in human history. It is a story of bondage and liberation, of divine signs and human struggle, of a people pulled from oppression into a wilderness they do not yet understand. And at its center stands Moses—the reluctant prophet turned liberator—bearing a message not only to Pharaoh, but to all future systems of power: **"Let my people go."**

With Aaron at his side, Moses returns to Egypt to confront Pharaoh, the very embodiment of earthly dominance. What follows is a clash not simply between two men or two nations, but between two visions of reality:

- One that clings to control, hierarchy, and fear
- And one that responds to suffering with sacred purpose

The plagues that unfold—water turned to blood, swarms of locusts, darkness over the land—are dramatic signs. But they are more than supernatural events; they are **disruptions of false order**. Each plague strikes at the heart of Egyptian life and religion, revealing the fragility of an empire built on domination.

When Pharaoh finally yields—after the most devastating sign, the death of the firstborns—the Israelites depart in haste. They do not walk into paradise. They walk into the **unknown**.

The parting of the **Red Sea** is perhaps the most iconic miracle in the Hebrew Bible. With Pharaoh's army pursuing, the waters open before them. The people cross on dry ground. When the Egyptians follow, the sea returns, swallowing the might of empire in a single gesture.

This is not just a historical claim—it is a **spiritual metaphor**: that true liberation requires faith in what cannot be seen, trust in what has not yet come, and the courage to move forward even when the path has not yet opened.

Yet freedom does not arrive all at once. What lies beyond the Red Sea is not immediate peace, but **wilderness**.

In the desert, the Israelites wrestle with fear, hunger, and longing for the familiarity of bondage. "Would that we had died in Egypt," they cry. It is easier to leave a place than to leave a mindset.

God responds with **manna from heaven**, **water from rocks**, and **guidance by fire and cloud**. These signs are not merely miracles—they are acts of relationship. A reminder that the same Presence that called Moses at the bush now walks with an entire people.

But the journey is not linear. The wilderness becomes a place of testing, purification, and transformation. It is where external slavery is exchanged for **internal formation**—where laws are given, identities are shaped, and hearts are refined.

The Exodus is not just about escape from Egypt. It is about becoming ready to receive something greater.

In the spiritual life, this pattern is universal:

- We are **called** out of what confines us
- We are **led** through what strips us down
- We are **prepared** to receive a truth we could not yet hold

The Exodus is not over. It continues wherever people rise from oppression, wherever courage challenges injustice, and wherever faith carries us through the wilderness within.

Queries for Reflection

- What "Egypt" have you had to leave behind in your own life?
- Can wilderness be not a punishment, but a place of preparation?
- What miracles might we miss when we long too quickly for certainty?
- How does the story of the Exodus echo in struggles for justice today?

5.4.4
Mount Sinai and the Revelation of Law

After the crossing of the sea and the long march into wilderness, the Israelites arrive at the foot of **Mount Sinai**—a jagged peak in a barren landscape, quiet yet trembling with mystery. It is here, not in a throne room or temple, that the next phase of the journey begins: the transformation of a scattered people into a **covenant nation**.

At Sinai, the Divine does not remain hidden. The mountain shakes. Thunder roars. Fire descends. A **voice speaks**, not to Moses alone, but to an entire community gathered below. For the first time in recorded tradition, **an entire people is invited into sacred dialogue**.

What follows is the giving of **the Law**—not merely as a list of rules, but as a framework for a holy life. The most famous of these are the **Ten Commandments**, inscribed on stone, offered as a covenant between God and the people. They speak to both vertical relationship (with God) and horizontal relationship (with one another):

- You shall have no other gods before Me
- You shall not make idols
- You shall not take the name of the Lord in vain
- Remember the Sabbath and keep it holy
- Honor your father and mother
- Do not murder
- Do not commit adultery
- Do not steal
- Do not bear false witness
- Do not covet

These are not arbitrary demands. They are invitations to **right relationship**—with the Divine, with community, and with the self. They seek not just to regulate behavior, but to **shape character**.

In a time and region marked by empire and inequality, the Law given at Sinai stands apart. It makes no distinction between rich and poor when it comes to justice. It calls for rest—not only for humans, but for animals and even the land. It sets rhythms of generosity, forgiveness, and ethical responsibility.

In this light, the Law becomes more than religious code—it becomes a **spiritual blueprint for society**.

And yet, even in the presence of revelation, doubt and disorder arise. While Moses lingers on the mountain in communion with God, the people below grow restless. They melt their gold and fashion a **golden calf**, a visible god they can control and touch. When Moses returns and sees what has happened, he is devastated. The tablets are shattered. Trust is broken.

This moment reveals something essential: that **revelation alone is not enough**. Even in the presence of miracles, the human heart struggles to remain faithful. The path of transformation requires more than awe—it requires continuity, commitment, and often, forgiveness.

Still, the covenant is renewed. Moses climbs again. A new set of tablets is given. The journey continues, not because the people are perfect, but because **the Divine is patient**.

Sinai becomes a symbol of both glory and tension—a meeting place of heaven and earth, of silence and speech, of law and love.

Queries for Reflection

- What does it mean to receive "law" not as control, but as sacred structure?
- Can limits be a form of freedom when rooted in wisdom and compassion?
- How do we respond when, even after insight, we fall into old patterns?
- What mountains in our own lives might still be waiting for us to climb—and listen?

5.4.5
The Prophets: Voices of Justice, Warning, and Hope

After Sinai, the story of Israel continues through kings, conquests, exile, and return. But woven through these centuries are the voices of the **prophets**—figures who stood between heaven and earth, speaking with fire, tears, and uncompromising clarity.

The prophets were not fortune-tellers. They were **truth-tellers**. They did not speak to satisfy the powerful, but to challenge them. They arose when kings strayed, when injustice spread, when rituals became hollow, and when the covenant was forgotten.

Their words were not always welcome. But they were necessary.

What Is a Prophet?

In Hebrew, the word for prophet is *navi*—one who is "called" or "brought forth." The prophet was not self-appointed. Their authority came not from position, but from a profound sense of being **spoken through**.

They were not mystics removed from society. They were deeply embedded in it—crying out in public squares, confronting rulers, comforting the exiled, and demanding that worship be matched by action.

To the prophets, religious observance meant nothing if it was not accompanied by **justice**.

"I desire mercy, not sacrifice," says Hosea.

"Let justice roll down like waters, and righteousness like a mighty stream," proclaims Amos.

"What does the Lord require of you," asks Micah, "but to do justice, to love kindness, and to walk humbly with your God?"

Judgment and Compassion

Prophetic messages often began with warning. They called out corruption, greed, idolatry, and the exploitation of the poor. But they did not speak only of judgment. Always, within their words, was a thread of **hope**.

They envisioned a time when swords would be beaten into plowshares. When the lion would lie down with the lamb. When hearts of stone would become hearts of flesh.

In the voice of **Isaiah**, we hear cosmic poetry and messianic longing. In **Jeremiah**, we hear weeping and lament—yet also promises of renewal. In **Ezekiel**, we see wild visions—wheels of fire, dry bones returning to life—symbols of transformation against all odds.

These prophets kept the flame of the covenant alive, even when the temple was destroyed, even in exile, even in silence.

They taught that God is not confined to a building or a ritual—but is found wherever justice is done, the widow is remembered, and the stranger is welcomed.

The Prophetic Legacy

The voice of the prophet did not end in antiquity. It echoes in every age and tradition where truth speaks to power. Where conscience rises against comfort. Where those with no voice are lifted by those who refuse to stay silent.

From abolitionists to civil rights leaders, from mystics to activists, the prophetic spirit continues to stir.

But it also stirs within.

Each of us, in moments of clarity, carries a spark of that same fire—the ability to speak with integrity, to act with courage, and to hold fast to hope in dark times.

Queries for Reflection

- What voices in our world today carry the prophetic spirit?
- Is it possible to balance critique with compassion, truth with love?
- How do we respond when religion serves power rather than justice?
- What does it mean to be a prophet—not with grand gestures, but with everyday integrity?

5.4.6
Mysticism in the Hebrew Tradition

Beneath the laws and the prophets, beyond the histories and the psalms, flows a quieter current in the Hebrew tradition—one of **mystical encounter**, hidden meanings, and the search for divine union. Often overlooked beside the grandeur of Exodus and the passion of the prophets, **Jewish mysticism** reveals a dimension of the tradition that is contemplative, symbolic, and vibrational.

It is a tradition not of spectacle, but of **depth**. Not focused on miracles, but on the mystery behind all things.

The Roots of Hebrew Mysticism

Mystical elements can be found even in the early scriptures. The encounter of Moses with the **burning bush** is not just prophetic—it is mystical: a moment of timelessness, a voice speaking from flame, a Name that is not a name but a riddle of being.

Later, **Elijah** meets God not in the earthquake or fire, but in a **still small voice**. The Hebrew word used—*demamah dakah*—suggests not just silence, but a silence that speaks.

These moments hint at a dimension of the Divine that is not loud or literal, but **intimate and inward**.

Kabbalah and the Tree of Life

In the medieval period, particularly in 12th–13th century Spain and southern France, Jewish mysticism blossomed into a structured tradition known as **Kabbalah**.

Kabbalah teaches that behind the visible world lies a realm of divine energies—**emanations** called *sefirot*—which flow from the Infinite (*Ein Sof*) into creation. These ten *sefirot*, arranged in the form of the **Tree of Life**, represent both aspects of God and aspects of the human soul.

- *Chesed* (Lovingkindness)
- *Gevurah* (Strength or Judgment)
- *Tiferet* (Beauty)
- *Yesod* (Foundation)
- *Malkhut* (Kingdom)

  ...and others, each reflecting an aspect of reality in dynamic relationship.

The goal is not merely to understand this structure, but to **live in harmony with it**—to participate in the healing of the world (*tikkun olam*) by aligning one's actions, speech, and thought with divine flow.

This mystical vision is not escape from the world, but deeper engagement with it.

The Zohar and Sacred Light

The **Zohar**, the foundational text of Kabbalistic literature, presents a mystical interpretation of the Torah. Written in Aramaic and filled with symbolic language, it sees the biblical stories as veils over deeper spiritual truths.

Adam and Eve are not just the first humans—they represent archetypal aspects of the soul. The exile from Eden is not just about geography, but about spiritual fragmentation. The Exodus becomes a journey not only from slavery to freedom, but from **separation to wholeness**.

In this view, the Torah is not merely a historical document, but a **living revelation**—one that must be read with the eyes of the heart as well as the intellect.

Mysticism as Inner Covenant

Where the prophetic path calls us to speak and act, the **mystical path calls us to listen and behold**. It is rooted in prayer, meditation, and the experience of unity with the Divine.

For some mystics, the ultimate goal is not reward or even understanding, but **closeness**—to become like Moses, who "spoke with God face to face," or like the Psalmist, who sang, "My soul thirsts for You… as a dry land thirsts for water."

This longing is not weakness. It is the beginning of wisdom.

Queries for Reflection

- What is the role of mystery in a world that favors certainty?
- Can sacred texts be read not only for instruction, but for revelation?
- How might we live more attuned to the unseen patterns of divine flow?
- In your own life, where have you felt the quiet, burning presence that words cannot name?

5.4.7
Moses and the Prophets in Interfaith Perspective

Moses and the prophets are central not only to Judaism, but to a **wider spiritual inheritance** shared across the Abrahamic faiths—and even beyond. Their voices, visions, and values ripple through Christianity, Islam, and global spiritual consciousness, serving as bridges between traditions and catalysts for conscience.

In every age and culture, the archetype of the prophet remains relevant: the one who hears a deeper call and dares to speak, not from self-interest, but from alignment with something greater.

Moses in Christianity

In Christian tradition, Moses is revered as **the great lawgiver**, a towering figure who delivers God's commandments and leads the Israelites from bondage. His role is foundational—not only historically, but spiritually.

The Gospel of Matthew presents Jesus as a **new Moses**, offering teachings from a mountaintop, much like Sinai. The Sermon on the Mount echoes and expands upon the moral vision first revealed in the Torah.

Yet Christian theology often frames Moses in contrast to Christ: law versus grace, old covenant versus new. While this framing has historically caused division, many contemporary Christian thinkers seek **continuity**—recognizing Moses not as replaced, but as fulfilled in the person of Jesus, who himself honored and quoted the Hebrew scriptures.

Moses in Islam

In Islam, **Musa (Moses)** is one of the most frequently mentioned prophets in the Qur'an—his story told with great detail, reverence, and moral weight.

He is a messenger, a liberator, a receiver of scripture (*Tawrat*), and a spiritual exemplar. His encounter with Pharaoh, his confrontation with injustice, and his moments of fear and doubt are all seen as profoundly human and deeply instructive.

In the Islamic mystical tradition (Sufism), Moses' dialogue with God on the mountain becomes a model of **intimate proximity**, and his longing to see the Divine (*"Let me behold You"*) is echoed in the heart of every seeker.

Moses is not just a prophet for the Israelites—he is a universal figure of courage, leadership, and submission to the will of God.

The Prophets Across Traditions

The Hebrew prophets are also honored in Christianity and Islam, not as relics, but as **living voices** whose messages remain essential.

- **Isaiah**, with his vision of peace and justice, is quoted throughout Christian liturgy.
- **Jeremiah**, the weeping prophet, is seen as a symbol of spiritual endurance.
- **Jonah**, whose story is found in all three traditions, reminds us of mercy, repentance, and the limits of self-righteousness.

These shared stories form a kind of **spiritual DNA** between Judaism, Christianity, and Islam—each interpreting the figures differently, yet recognizing in them a profound transmission of divine concern for justice, humility, and transformation.

Beyond the Abrahamic Traditions

While Moses and the Hebrew prophets are most prominent within the Abrahamic world, their themes resonate far beyond.

- The call to **liberation** echoes the teachings of the Buddha and the path of inner freedom.
- The emphasis on **right action** and covenant aligns with the *dharma* of Hinduism and the ethical codes of Confucianism.
- The prophetic critique of power and wealth is mirrored in Indigenous wisdom and modern social movements alike.

In this way, Moses and the prophets speak not just to the faithful, but to the **human spirit itself**—wherever there is suffering, silence, or longing for justice.

Queries for Reflection

- How do different traditions shape and share the memory of Moses and the prophets?
- Can a prophetic voice belong to multiple faiths at once—and still speak freshly to all?
- What does it mean to be a vessel of truth in a world of noise and fear?
- Might the prophets still be calling—not from ancient scrolls, but from within us?

5.4.8

Following the Flame: Calling, Conscience, and Covenant

From the burning bush to the blazing words of the prophets, the Hebrew tradition burns with one unextinguished flame: the call to **live in covenant**—not just with God, but with justice, truth, and one another.

This covenant is not a contract. It is a **relationship**—fierce, dynamic, and alive. It asks more than belief. It demands **participation**. It does not call the perfect. It calls the willing.

Moses, who stuttered and doubted, became the liberator of his people. The prophets, who suffered and lamented, became the conscience of their generation. None were chosen because they were flawless. They were chosen because they **listened**—and responded.

To follow this flame is not to mimic their lives, but to enter their **spiritual posture**:

- To turn aside when something burns and speaks
- To walk into conflict with humility and courage
- To speak even when the words tremble
- To act even when the path is uncertain

The prophetic tradition reminds us that **faith is not passive**. It is not merely private. It is expressed in how we treat the widow and the stranger, how we respond to injustice, how we remember those the world forgets.

And yet, at its core, it is also **personal**. It begins in the quiet places—where we hear the still small voice, where we wrestle with our own resistance, where we ask, like Moses: *"Who am I?"*

The answer, then and now, is the same: *You are the one who was called. The one who turned aside. The one who said yes.*

You do not have to be a prophet in title. To live with conscience is to walk in their footsteps.

You do not have to hear a voice from a bush. Sometimes the call comes as a whisper in your own heart.

You do not need to part seas. But you may be asked to help someone else cross.

To follow the flame is to live as if your life matters—because it does.

It is to remember that the story is still unfolding. And that the covenant, ancient as it is, has never been closed.

Final Queries for 5.4

- What does it mean to be called—not once, but continually?
- How do we respond when conscience and comfort come into conflict?
- What would it look like to renew our own covenant—with truth, with justice, with life itself?
- Are we willing to follow the flame, even when it leads into the unknown?

5.

MUHAMMAD AND MESSENGER

Seal of the Prophets, Bearer of the Word

-ZEPHYRO SOLON

CONTENTS

5.5.1
The World Before the Revelation: Arabia in the 6th Century

Before the revelation of the Qur'an, before the rise of a new spiritual path, the world of Muhammad was already in motion—vast, tribal, poetic, and complex. To understand his emergence as a prophet, we must first glimpse the **landscape of pre-Islamic Arabia**, a world both rich in tradition and aching for transformation.

A Land of Tribes and Traditions

6th-century Arabia was a desert region of scattered oases, trading towns, and nomadic tribes. There was no single ruler, no centralized state—only a tapestry of clans bound by blood, honor, and survival. Each tribe held its own customs, alliances, and internal codes of justice.

At the center of this world stood **Mecca**, a thriving commercial hub and a place of spiritual significance. It housed the **Kaaba**, a cube-shaped sanctuary believed to be built by Abraham and Ishmael, though by Muhammad's time it had become a center for **polytheistic worship**, with idols representing many deities.

Pilgrimages to the Kaaba brought both commerce and ritual. The surrounding poetry, marketplaces, and oral traditions reflected a vibrant, if fragmented, culture.

The Social and Moral Climate

While tribal codes promoted loyalty and generosity within clans, the broader social fabric was marked by inequality and conflict. Blood feuds could last generations. The weak—including women, orphans, and slaves—were often left vulnerable. Female infanticide, though not universal, was practiced by some tribes.

Religious life varied widely. Some Arabs worshipped multiple gods, others were loosely affiliated with Judaism, Christianity, or ancient monotheistic traditions. But there was no unified theology, and **many felt spiritually adrift**.

Among this diversity were seekers known as **Hanifs**—monotheists who rejected idolatry and sought the pure worship of one God, drawing from echoes of the Abrahamic tradition. Muhammad is said to have been among them.

A World Ready for Revelation?

While much of the ancient world had settled into empires, Arabia remained tribal, decentralized, and spiritually open. In this space—untethered to the power structures of Rome or Persia—something new could emerge.

Muhammad's world was not lacking in religious ritual or poetic expression. What it lacked was **cohesion**—a unifying vision that could transcend tribalism, uplift the marginalized, and root everyday life in a deeper moral and spiritual order.

Into this setting—restless, rhythmic, and ready—a quiet man from the Quraysh tribe would one day retreat to a cave in search of solitude.

And from that cave, a voice would rise that would echo across the world.

Queries for Reflection

- How does a spiritual revolution begin—not with power, but with longing?
- What conditions—social, moral, or existential—prepare the way for revelation?
- Can fragmentation and diversity be the soil from which unity and purpose arise?
- What deserts of our own time might be awaiting a new voice?

5.5.2
The Cave at Hira: The First Revelation

Muhammad was not born a prophet. He was born an orphan, into the respected but not powerful **Quraysh tribe** in Mecca. As a child, he experienced both hardship and protection, growing up under the care of his grandfather and then his uncle. Known for his honesty and integrity, he was called **al-Amin**—"the trustworthy one." As he matured, he became a merchant, married the noble and wise **Khadijah**, and began to seek **solitude**.

Even before revelation, Muhammad was a **seeker**.

He would retreat often to the **Cave of Hira**, a small hollow in the mountains just outside Mecca. There, amid the silence and stone, he would contemplate existence, the injustices of his society, and the nature of the divine.

It was during one of these retreats—**in the month of Ramadan**, around the year 610 CE—that something happened which would change not only his life, but the spiritual trajectory of the world.

According to tradition, an angel—later identified as **Jibril (Gabriel)**—appeared in the cave and **commanded him to recite**:

"*Iqra!*"

Read! or *Recite!*

Muhammad, startled and afraid, responded that he could not read. Again the command came. And again. Until, finally, the words emerged—burning with clarity, carried by the unseen:

"Recite in the name of your Lord who created—
Created man from a clinging substance.
Recite, and your Lord is the Most Generous—
Who taught by the pen—

Taught man that which he knew not."
—(Qur'an 96:1–5)

The moment was overwhelming. He fled the cave in fear, trembling and unsure of what had occurred. He returned to Khadijah, who comforted and reassured him. She brought him to her cousin **Waraqah ibn Nawfal**, a Christian mystic familiar with previous revelations, who confirmed: *This is the same Spirit that came to Moses.*

A Reluctant Prophet

Like many before him, Muhammad did not seek prophethood. His first response was not triumph, but awe. Yet the revelations continued. Over the next 23 years, through triumphs and trials, battles and betrayals, the **Qur'an** would descend—not all at once, but in verses and chapters, shaped by events and moments, memorized and recited by heart.

What began in a cave—a man alone with his questions—would become the foundation of a global civilization.

But it began with **vulnerability**. With trembling. With surrender.

Muhammad's prophetic call was not about self-glorification. It was about becoming **a vessel**—transparent enough to carry a message larger than himself.

His earliest followers were few: his wife, his friend Abu Bakr, his young cousin Ali, his servant Zayd. But they were devoted not to his charisma, but to the **truth** that flowed through him.

Revelation, in this tradition, is not imposed from above, but **received in the depth of listening**.

Queries for Reflection

- What does it mean to be called when one feels unprepared?
- How does revelation arise—not in crowds, but in caves?
- Can fear and uncertainty be part of the path to transformation?
- Where in your life are you being asked to listen more deeply?

5.5.3
The Qur'an: Message and Meaning

The **Qur'an** is not merely a book—it is, for Muslims, the **literal word of God**, revealed through the Prophet Muhammad over 23 years. Its words were not authored, edited, or composed by the Prophet; they were **received**—sometimes in moments of stillness, sometimes in the midst of crisis—and then recited aloud, committed to memory, and eventually written down.

Even the word *Qur'an* means "recitation." Its rhythm, its sound, and its **spoken quality** are essential to its nature. More than a text, the Qur'an is a **living voice**, meant to be heard as well as read.

A Message for Its Time—and Beyond

The Qur'an addresses both **eternal themes** and **immediate realities**. It speaks of the unity of God (*tawḥīd*), the purpose of creation, and the Day of Judgment. It also responds to specific events in Muhammad's life: battles, betrayals, treaties, and community disputes.

Yet always, it returns to the same spiritual core:

- God is **One**, beyond image, birth, or likeness.
- Human beings are **stewards** on Earth, entrusted with free will and moral choice.
- Every action is recorded. Every soul will be called to account.
- God is **merciful and just**, nearer than your jugular vein, yet beyond comprehension.

The Qur'an calls its readers to **reflect**. Again and again it asks:

"Will you not think?"
"Will you not use your reason?"
"This is a sign for those who understand."

It offers parables, reminders, stories of earlier prophets—Adam, Noah, Abraham, Moses, Jesus—not to entertain, but to awaken.

Law, Guidance, and Mercy

The Qur'an contains moral and legal guidance, but it is not a legal code. Its spirit is one of **balance**—between individual conscience and communal responsibility, between ritual worship and social justice.

It teaches prayer, fasting, charity, and pilgrimage. But just as strongly, it condemns exploitation, injustice, arrogance, and neglect of the poor.

It speaks not only of duties toward God, but of duties toward **neighbors, orphans, spouses, strangers, and even enemies**.

"We have sent you not but as a mercy to the worlds."
—(Qur'an 21:107)

This verse, addressed to Muhammad, reveals the underlying ethos: that the Qur'an's purpose is not domination, but **compassion**.

Beauty and Power

For those fluent in Arabic, the Qur'an's language is often described as **inimitable**—its style unmatched, its cadence moving. Even early skeptics in Mecca admitted its force. Many converts began as opponents, only to be overcome by the *sound* and *spirit* of what they heard.

Even in translation, one can sense its layered beauty—its sudden shifts between voice and address, its use of cosmic imagery, its intimate and commanding tone.

"By the dawn and the ten nights…
By the sun and its brightness…
By the fig and the olive…

By time—truly, humanity is in loss…"
—*(Selections from Surahs 89, 91, 95, 103)*

The Qur'an does not argue—it **reveals**. It does not persuade by force, but by **evocation**. It does not merely inform—it **invites**: to reflect, to remember, to return.

Queries for Reflection

- Can scripture be both time-bound and timeless—responding to history yet pointing beyond it?
- What does it mean to hear words as revelation, not just information?
- How might the rhythm and beauty of a sacred text carry meaning beyond translation?
- What messages might the Qur'an offer not only to Muslims, but to the wider human family?

5.5.4
Prophet and Statesman: Building a Community

Muhammad's mission was not only spiritual—it was **civilizational**. From a solitary seeker in a cave, he became a leader tasked with guiding a growing community through persecution, migration, conflict, and unification. He was not simply a messenger of divine truth; he became a **builder of society**, embodying the Qur'an not only in word, but in action.

This transformation did not happen overnight. It unfolded through decades of struggle and discernment—first in **Mecca**, then in **Medina**, as the early Muslim community grew and faced both internal and external challenges.

The Meccan Years: Message in the Face of Persecution

In Mecca, Muhammad's message was **radical**: that there is one God above all tribal deities; that the wealthy have responsibilities to the poor; that human beings are equal before their Creator. These teachings challenged the commercial and social power structures of the Quraysh elite.

As a result, the early Muslims faced **persecution, mockery, and violence**. Some were tortured. Others fled. Yet the Prophet remained patient, calling his followers to stand firm in faith, to return hatred with kindness, and to rely on the mercy of God.

His leadership in Mecca was moral and spiritual—but not yet political.

The Hijra: A New Chapter in Medina

Everything changed with the **Hijra**—the migration to **Yathrib**, later known as **Medina**. There, Muhammad was invited not just as a preacher, but as an **arbiter** between rival tribes. It was in Medina that his role as statesman emerged.

He drafted the **Constitution of Medina**, one of the earliest known written agreements to recognize religious pluralism and collective responsibility. It established Muslims, Jews, and other groups as members of a single community (*ummah*), bound by mutual protection and justice.

In this role, Muhammad became a judge, peacemaker, military strategist, and administrator—**a prophet with dusty sandals**, navigating real-world complexities without compromising the ethical core of his message.

War and Restraint

In Medina, Muhammad and his followers were drawn into armed conflict, largely in defense. The Qur'an permitted fighting to protect the community, but emphasized ethical boundaries:

"Fight in the cause of God those who fight you, but do not transgress. God does not love the aggressors."
—(Qur'an 2:190)

Rules were established: no harm to civilians, no destruction of crops, no killing of monks. Even in war, restraint was part of revelation.

The Prophet's character during these years revealed a **deep balance** between resolve and mercy. When Mecca was finally reclaimed years later, he did not seek revenge. He offered a **general amnesty**, forgiving many of those who had once tried to kill him.

Leading by Example

Muhammad lived simply. He patched his clothes, served in his home, and walked among his people. He discouraged flattery and refused luxury. He laughed, cried, counseled, prayed. Those closest to him described him as **"the Qur'an walking."**

His leadership was not driven by dominance, but by **trust**—earned through consistency, humility, and devotion to the greater good.

In Muhammad, we see a rare synthesis: spiritual depth and practical wisdom, revelation and governance, contemplation and compassionate authority.

Queries for Reflection

- Can spiritual leadership extend into civic life without losing its soul?
- What does it mean to lead not through force, but through example?
- How do we balance justice and mercy in positions of responsibility?
- In what ways might Muhammad's leadership model offer insight for our own divided world?

5.5.5
The Night Journey and the Inner Path

Among the most mysterious and beloved moments in the Prophet Muhammad's life is the **Night Journey and Ascension**—*al-Isrāʾ wa al-Miʿrāj*—an event that defies conventional categories of time, space, and understanding.

According to tradition, in the depth of night, the Prophet was transported from **Mecca to Jerusalem**, and from there ascended through the **heavens**, meeting earlier prophets, witnessing signs of the unseen realms, and approaching the divine presence itself.

This event, while described in both literal and mystical terms, has come to symbolize the **inner dimension** of the prophetic path—the journey not only across lands, but **into the soul**.

The Journey Begins: From the Sacred Mosque to the Farthest Mosque

The Qur'an briefly references the event:

"Glory be to Him who took His servant by night from the Sacred Mosque to the Farthest Mosque, whose surroundings We have blessed…"
—(Qur'an 17:1)

This first leg—*al-Isrāʾ*—takes Muhammad from the **Kaaba in Mecca** to **al-Aqsa in Jerusalem**, linking the Islamic tradition to the lineage of earlier prophets: Abraham, Moses, Jesus. It is a journey of **continuity**, placing Muhammad within the same stream of divine guidance that flowed through the Hebrew and Christian scriptures.

In Jerusalem, he is said to have led them in prayer—a gesture not of supremacy, but of spiritual **solidarity**.

The Ascent: Beyond the Veils

From there, he ascends—*al-Miʿrāj*—through seven heavens, meeting Adam, Jesus, Joseph, Idris, Aaron, Moses, and Abraham. Each encounter affirms his mission and connects him with the **chain of prophecy**.

At the highest point, beyond even the angel Gabriel, the Prophet is said to come **face to face with the Divine**—in a realm known as the **Lote Tree of the Utmost Boundary** (*Sidrat al-Muntahā*). Here, he receives the command for the **daily prayers** (*salat*), which would become the central ritual of Islamic practice.

Though interpretations vary, this moment signifies a **direct experience of divine presence**—not mediated by text or tradition, but through soul-to-soul encounter.

A Journey of the Heart

For many Muslims, the Night Journey is both a historical miracle and a **symbolic map of the soul**. It reflects the **spiritual ascent** available to all: from outer discipline to inner awakening, from worldly distraction to sacred nearness.

Just as the Prophet journeyed upward through layers of being, the seeker is invited to ascend through **prayer, remembrance, and surrender**. In this view, the journey is not only his—it is **ours**.

The five daily prayers—received in the heart of that night—become more than obligation. They are the **echo of ascent**, a chance to return, step by step, to presence.

The Return to Earth

Significantly, Muhammad does not remain in the heights. He returns—to his community, to the dust and labor of daily life. His night of union leads not to withdrawal, but to deeper engagement.

In this, we see the pattern of the prophetic path: **ascend to the Real, then descend with compassion**. Seek truth, then live it. Touch the infinite, then walk among the finite with grace.

Queries for Reflection

- What does it mean to ascend—not physically, but spiritually?
- Are we willing to be led beyond our understanding into the unknown?
- How do we integrate mystical experience with everyday responsibility?
- Can prayer become more than ritual—can it become return?

5.5.6
Mysticism and Inner Dimensions

While many encounter Islam through its outward expressions—ritual prayer, ethical law, and communal practices—there is also a rich inner dimension that points toward the heart of the faith: union with the Divine. This mystical path is most fully embodied in **Sufism**, a tradition that seeks not only to follow the Prophet Muhammad outwardly but to become inwardly transformed by love of God.

Sufism is not a separate sect but a stream that runs through Islamic tradition, touching both Sunni and Shia communities. Its practitioners, the **Sufis**, describe the journey toward God as a movement from the ego-self *(nafs)* toward the Divine Presence *(al-Haqq)*. This path involves purifying the heart, remembering God *(dhikr)*, and cultivating a deep, experiential awareness of divine nearness.

The Prophet as Mystic

The life of the Prophet Muhammad is seen by many Sufis not only as the model of action, but also of contemplation. His retreats to the cave of Ḥirā' before his revelations began, his night journey and ascension *(Miʿraj)*, and his reported spiritual states all point toward a dimension of inward intimacy with the unseen.

He is often described in Sufi poetry and literature as the "Beloved of God," the archetype of perfected humanity, and the one who most fully embodies the Light of God *(nūr Muhammadī)*. In this view, the Prophet is not worshipped, but deeply loved—as the bridge between the finite and the infinite.

The Path of Love

For many mystics, the central force that draws the soul toward God is **love**. The Sufi poets—Rumi, Hafez, Attar, and others—speak of love not as sentiment but as fire: a burning away of all that is false, a longing that purifies the self and prepares it for union.

This love is not always peaceful. It can be overwhelming, even painful. But it is this intensity that brings about transformation. God is not only the Judge or the King, but the Hidden Treasure who longs to be known. And the human being, in the Sufi worldview, is not merely a servant, but a lover seeking reunion.

Practices of the Heart

Sufi orders (*ṭuruq*) each have their own methods and teachings, but many share a common set of inner practices. These may include:

- **Dhikr (Remembrance):** Repeating divine names or phrases to focus the heart.
- **Sama' (Listening):** Engaging with poetry or sacred music to awaken longing.
- **Muraqabah (Watchfulness):** A meditative awareness of God's presence.
- **Whirling or Movement:** In some traditions, physical motion becomes a form of devotion.

These practices are not ends in themselves. They are tools for awakening—to shift from knowing about God to knowing God directly.

The Inner and the Outer

Sufism emphasizes that outer practice (*sharī'ah*) and inner realization (*ḥaqīqah*) are not opposites. Rather, they form a complete path when held together. The law provides structure; the spirit gives it life.

This integration echoes one of the Prophet's reported sayings: "*The first knowledge to be lifted from the world is inner knowledge.*" Mysticism, then, is not deviation—it is remembrance. A return to the heart of the matter.

5.5.7
Mysticism and Devotion in Islam

Beneath the surface of Islamic law and theology flows a river of **mystical longing**—a current of devotion that seeks not only to obey God, but to **know God intimately**. This stream is known broadly as **Sufism**, or **taṣawwuf**—a tradition rooted in the Qur'an and the life of the Prophet, yet reaching beyond form into **union, love, and surrender**.

Mysticism in Islam is not a fringe. It is a **heart within the body**—an inner path walked by scholars and poets, shepherds and kings, men and women, from the deserts of Arabia to the shrines of Persia, the forests of Africa, and the streets of Istanbul.

The Prophet as Mystical Model

The Prophet Muhammad is not only the transmitter of revelation—he is also the **first mystic** of Islam. His night vigils, long prostrations, moments of tears and silence, and his direct experiences with the divine presence (such as the *Miʿrāj*, the Night Ascension) all form the **spiritual blueprint** for later seekers.

He prayed not from obligation alone, but from intimacy. He said,

"My comfort is in prayer."

He would stand through the night, his feet swollen from devotion, not seeking favor but **proximity**.

To love God, he taught, is to love neighbor, to forgive, to soften the heart. The mystical path in Islam finds its **root in his tenderness**.

The Way of the Sufis

Sufis often describe the journey as passing through **three stages**:

1. **Shariah** – the outer path of ethics and law
2. **Tariqah** – the inner path of discipline and purification
3. **Haqiqah** – the realization of truth and nearness to the Divine

This is not a rejection of structure but a movement **through it**, from outer practice to inner transformation.

Through **zikr** (remembrance), **sama** (spiritual listening), **poetry**, and **service**, the seeker polishes the heart until it becomes a mirror reflecting the One.

"Die before you die," the Sufis say—meaning, let the ego fall away before the body does.

The Language of Love

No discussion of Islamic mysticism is complete without mentioning its great **poets and saints**—those who gave language to the inexpressible.

- **Rumi**, whose verses circle like the dervishes who dance in his name, speaks of the soul's longing for reunion with its Source:

  "I am not this hair, I am not this skin,
  I am the soul that lives within."
- **Rabia of Basra**, an early woman mystic, offered her devotion without seeking heaven or fearing hell, saying:

  "I do not want heaven or hell. I want only You."
- **Al-Hallaj**, who cried "*Ana al-Haqq*" (*I am the Truth*), was executed for uttering what mystics knew: that in the depth of realization, the self disappears and only the Divine remains.

These mystics do not speak of God as an idea—but as a **Beloved**, closer than breath, more real than thought. For them, every moment is an invitation to love, and every hardship a veiled form of grace.

Mysticism as Mercy

In a world often torn by division, mysticism offers **bridges**—between reason and heart, between traditions, between human and divine.

Sufism, though distinctively Islamic, has always been **open to the beauty in others**. Many Sufis studied Christian, Jewish, Hindu, or even Greek philosophical texts—not to dilute the message, but to enrich it.

This openness arises not from relativism, but from a **deep certainty** that truth cannot be contained by one form alone.

Queries for Reflection

- What does it mean to love the Divine—not from fear, but from longing?
- How might spiritual practice become a path of inner transformation, not just outer conformity?
- Can mysticism bridge the gap between traditions by focusing on essence over form?
- What might happen if prayer became not duty, but dialogue?

5.5.8
Muhammad in Global and Interfaith Perspective

Today, Muhammad stands as one of the most influential figures in world history—revered by over a billion Muslims as the final prophet, yet often misunderstood or reduced to caricature outside the tradition he helped shape. To understand Muhammad not only as a religious figure but as a **global moral exemplar**, we must move beyond doctrine into **dialogue**.

He was a man of **deep prayer and practical leadership**, of **compassion and courage**, of **spiritual illumination and political challenge**. The world he entered was fragmented, tribal, and steeped in injustice. What emerged from his life was not only a sacred text, but a living civilization—marked by law, poetry, science, and soul.

Beyond Misconception

In many Western portrayals, Muhammad has been viewed through the lens of conflict. Critics have emphasized battles, legal rulings, or cultural differences while ignoring the full picture of his life and teachings.

Yet history tells a richer story:

- Of a man who wept in prayer and forgave his enemies.
- Who elevated women's rights in a patriarchal society.
- Who condemned racism, saying, "No Arab is superior to a non-Arab, nor a non-Arab to an Arab…"
- Who embraced Jews and Christians as "People of the Book," acknowledging the light in their scriptures.
- Who was praised by Gandhi for his commitment to justice and simplicity.

- And whose influence shaped not only religion, but **language, medicine, architecture, and ethics** from Spain to Indonesia.

To study Muhammad fairly is not to sanitize him, but to **see him fully**—as a bridge between heaven and earth, the eternal and the everyday.

Interfaith Resonance

In the Qur'an, Muhammad is called a **"mercy to all the worlds"** (*rahmatan lil-ʿālamīn*). His example has inspired not only Muslims, but seekers of all paths who recognize in him a life of principle, prayer, and profound humility.

While theological differences remain between Islam and other religions, **Muhammad's life creates openings**:

- For Jews, he upholds the prophetic tradition of justice, covenant, and community.
- For Christians, he honors Jesus not as rival, but as beloved messenger.
- For Hindus and Buddhists, his teachings on discipline, compassion, and ego transcendence mirror deep currents of their own wisdom.

In many interfaith circles, Muhammad is increasingly seen not as a barrier, but as a **bridge**—a figure who can invite respectful conversation, ethical inspiration, and shared values across traditions.

The Human Mirror

To Muslims, Muhammad is not divine. He is **fully human**—yet through his humanity, the light of the Divine shines. He is called **al-Insān al-Kāmil**, the "perfected human"—not in flawlessness, but in his alignment with truth, justice, and love.

His life remains **emulable**. Not distant, but close. Not an abstract legend, but a companion for those seeking to walk the path with sincerity.

In an age of disconnection and distraction, his presence reminds us of the power of stillness. The power of mercy. The power of listening—to God, to conscience, to each other.

Final Queries for 5.5

- How can we move beyond distorted images of religious figures and seek to understand them on their own terms?
- In what ways might Muhammad's life offer guidance for modern leadership, interfaith respect, or personal practice?
- What if prophecy was not about perfection, but about **presence**?
- How might we, like him, become messengers of mercy—however small our circle may be?

6.

LAOZI AND CONFUCIUS

Wisdom from the East: Flow and Order

-ZEPHYRO SOLON

CONTENTS

5.6.1
Ancient China: Harmony, Chaos, and the Search for Order

Long before the rise of centralized empires, the landscape of ancient China was shaped by a deep tension between **chaos and order**, instability and the longing for harmony. It was a world of competing warlords, fractured states, and philosophical ferment—a crucible of crisis that gave birth to some of the most enduring spiritual and ethical traditions the world has ever known.

By the 6th century BCE, China had entered what would later be called the **Axial Age**—a period when multiple civilizations across the globe gave rise to profound shifts in consciousness. In China, this was the time of the **Zhou dynasty**, during which the idea of the **Mandate of Heaven** took root: that rulers governed not by force alone, but by moral legitimacy. When they failed to uphold justice, they lost the right to rule.

But the reality on the ground was far from stable. The Zhou court weakened, local lords grew stronger, and the land descended into what would later be remembered as the **Spring and Autumn period**, and eventually the **Warring States period**—a time of near-constant conflict.

In the midst of this disarray, thinkers and teachers emerged—not to seize power, but to **seek wisdom**. They asked:

- How can a society flourish without falling into violence or corruption?
- What is the proper role of the ruler, the family, the individual?
- What does it mean to live in alignment with the natural order?

These questions were not academic—they were existential. And out of them arose two great and contrasting voices: **Laozi**, the sage of stillness, and **Confucius**, the sage of structure. One turned to the eternal rhythms of the cosmos. The other to the restoration of moral and social foundations.

Their teachings were not isolated philosophies, but answers to a shared human condition: the need to find a **path**—whether by surrendering to the Tao or by cultivating virtue through ritual and responsibility.

In a world unraveling, they offered two visions of how to **weave it back together**—one through **flow**, the other through **form**.

Queries for Reflection

- How does social disorder invite new forms of wisdom to emerge?
- Can harmony be found through both surrender and structure?
- What parallels exist between ancient China's crisis and our own time?
- When faced with a broken world, do we respond by retreating into stillness—or stepping forward into service?

5.6.2
Laozi and the Way of the Tao

Laozi—literally "Old Master"—is a name shrouded in mystery. Whether he was a historical figure, a collective voice, or a poetic legend, his influence is undeniable. Traditionally credited as the author of the *Tao Te Ching*, Laozi offered a vision of wisdom that does not command or construct, but **flows and dissolves**, like water wearing down stone.

The core of Laozi's teaching is the **Tao**—the Way. The Tao is not a god, nor a force to be worshipped. It is the **underlying principle** of reality itself. It gives rise to all things yet cannot be grasped. It cannot be named without losing its essence.

"The Tao that can be told is not the eternal Tao.
The name that can be named is not the eternal name."
—*(Tao Te Ching, Chapter 1)*

Laozi teaches that the more we try to control, the more we lose touch with the Way. Civilization, in its striving and structure, often cuts us off from the quiet rhythm that governs all life. The Tao, on the other hand, is **natural, spontaneous, and whole**.

To return to the Tao is not to escape the world, but to move **within it differently**. To act without forcing. To lead without dominating. To achieve without claiming.

This principle is called **wu wei**—often translated as "non-action," though it really means **effortless action** or **action in alignment**. It is not passivity, but presence. A tree bends in the wind and does not break. Water nourishes all things and seeks no praise. The sage, Laozi says, becomes like this—quiet, humble, effective without striving.

Laozi does not urge moral reform or political programs. He invites us to **unlearn**—to return to simplicity, softness, and stillness.

He warns against too much knowledge, too much desire, too much interference. "The more prohibitions there are," he writes, "the poorer the people become." He is not advocating chaos, but a return to **natural order**, one that cannot be engineered but only encountered.

The image of Laozi riding an ox westward, leaving behind the world of politics, symbolizes more than withdrawal. It represents a path that trusts the Tao to do what ambition cannot. It asks us to let go—not of responsibility, but of the illusion of control.

To follow Laozi is to look inward and around, to sense the pulse beneath all things, and to walk with a lighter touch.

Queries for Reflection

- Can wisdom arise through unlearning as much as through study?
- What would it mean to act without striving, to lead without force?
- How might our lives change if we trusted the natural unfolding of things?
- Is there a place for softness and stillness in a world of noise and speed?

5.6.3
The Tao Te Ching: Non-Action and Natural Flow

The *Tao Te Ching*, attributed to Laozi, is one of the most translated books in the world—second only to the Bible. Yet it remains, paradoxically, one of the most enigmatic. Comprising only 81 short chapters, it reads less like a manual and more like a whispered koan. There are no arguments, no doctrine, no linear logic—just a cascade of paradoxes, images, and inversions that seem to unravel the mind only to still it.

It opens with a warning: that the Tao which can be spoken is not the eternal Tao. From the outset, readers are disarmed. This is not a book meant to explain, but to **evoke**. It does not seek to inform so much as to realign.

The title itself can be loosely translated as *The Way and Its Power. Tao* (道) means way, path, or principle. *Te* (德) refers to inner virtue or character. *Ching* (經) means classic or scripture. Together, they point not to a fixed theology but to a **way of being** that flows in harmony with the underlying structure of the cosmos.

A Scripture of Paradox

The genius of the *Tao Te Ching* is that it speaks in opposites:
"The soft overcomes the hard."
"To yield is to be preserved whole."
"Those who know do not speak. Those who speak do not know."

This is not contradiction for its own sake. It is **poetic provocation**—a way of loosening the grip of the rational mind so something deeper can be sensed. Laozi does not provide answers; he dissolves the need for them.

In the Taoist vision, the world is not to be conquered or corrected. It is to be **understood and flowed with**. The sage does not seek to impose will, but to move with the natural current of things. This is the root of **wu wei**—the principle of non-action or effortless action.

Wu Wei: The Art of Effortless Effectiveness

Wu wei is one of the most misunderstood concepts in Taoism. It is not laziness, passivity, or inaction. Rather, it is **action without force**. It is the way a tree grows, a river winds, a mother nurtures—without striving, without contrivance, and yet accomplishing everything necessary.

The sage, says Laozi, is like water: yielding, lowly, yet vital and persistent. Water does not fight the stone, but wears it down. It nourishes all things and never asks for credit. Its power lies in its humility.

"The highest good is like water.
Water gives life to the ten thousand things and does not strive."
—*(Tao Te Ching, Chapter 8)*

In a world obsessed with ambition, assertion, and accumulation, wu wei offers a radical alternative: **ease, receptivity, and trust**. It asks us to become so attuned to the natural rhythm of life that our actions arise spontaneously, harmoniously, without friction.

This is not a retreat from responsibility. It is a refinement of presence.

Power Reimagined

The *Tao Te Ching* has been read by emperors and anarchists alike. One of its most enduring contributions is its **redefinition of power**.

Laozi warns against coercion, expansionism, and control. The more rules a society has, he observes, the more cunning and disobedience arise. The more weapons are hoarded, the more danger increases. True power, he suggests, is quiet, restrained, and invisible.

"A leader is best when people barely know he exists.
When his work is done, the people say, 'We did it ourselves.'"
—*(Tao Te Ching, Chapter 17)*

This ethic of **non-domination** is not only political—it's deeply spiritual. It speaks to a way of relating with life that is rooted in trust rather than fear, in wholeness rather than fragmentation.

Nature as Teacher

The *Tao Te Ching* draws its metaphors from **nature**. The Tao is not described, it is shown: in valleys, rivers, winds, seasons, and silence. Nature does not resist itself. It simply is. And in this is its beauty.

For Laozi, returning to nature is not a call to escape civilization, but to **remember simplicity**. Complexity, he warns, often leads to confusion. The Tao, like nature, is subtle and patient. It teaches not through commands, but through example.

In this sense, the *Tao Te Ching* is not a book to be mastered, but one to **become quiet enough to receive**.

A Companion for All Paths

Though deeply rooted in Chinese language and culture, the *Tao Te Ching* resonates with seekers across traditions. Its echoes can be found in Buddhist emptiness, Christian mysticism, Islamic Sufism, and indigenous reverence for the natural world.

Its enduring appeal lies in its **humility and openness**. It does not ask you to believe anything. It simply offers observations—gentle, paradoxical, disarming—and invites you to consider whether a softer way might be not only possible, but wiser.

A Text to Return To

No one "finishes" the *Tao Te Ching*. It reads differently each time, depending on where you are in life, what you're holding, and what you're willing to set down. Sometimes it is a mirror. Sometimes it is a guide. Sometimes it is a breeze passing through the mind, saying nothing—but changing everything.

Queries for Reflection

- How does effortless action differ from passive inaction?
- Where in your life are you striving—and what might happen if you let go?
- Can paradox be a doorway to deeper truth rather than confusion?
- What might leadership, love, and creativity look like when shaped by water, not fire?

5.6.4
Confucius and the Path of Virtue

If Laozi sought harmony through withdrawal and inner stillness, **Confucius** sought it through **engagement**—with family, society, tradition, and the cultivation of character. Where Laozi urged a return to the uncarved block, Confucius asked us to become **refined instruments**—shaped not by force, but by continuous practice of virtue.

Known in Chinese as **Kongzi** or **Kong Fuzi** ("Master Kong"), Confucius lived in the 6th century BCE during a time of widespread political disorder and cultural fragmentation. Like Laozi, he was responding to a civilization in crisis. But his solution was not to escape the world—it was to **repair it** from within.

The Life of the Teacher

Confucius was born into a minor noble family in the state of Lu. Though not wealthy, he received a classical education and eventually worked as a teacher, scholar, and minor government official. Disheartened by political corruption and disappointed in his attempts to serve as a reformer, he spent much of his later life traveling from state to state, offering his counsel to rulers and gathering a group of loyal students.

Unlike Laozi, who left behind only poetry, Confucius' words were recorded by his disciples in what became known as the **Analects**—a collection of sayings, conversations, and reflections that form the heart of **Confucian thought**.

His teachings were never meant to form a religion. They were meant to form **human beings**—capable of restoring balance, dignity, and harmony in a fractured society.

The Cultivation of Virtue (De)

Central to Confucianism is the concept of **virtue**—not as abstract moralism, but as embodied character. The aim is to become a **noble person** *(junzi)*—one who acts with sincerity, kindness, and integrity, regardless of recognition or reward.

Confucius taught that virtue must be **cultivated** through daily life—through relationships, habits, speech, and conduct. There is no shortcut to wisdom, only steady self-refinement.

He emphasized several key virtues:

- **Ren** (仁): humaneness, compassion, benevolence
- **Li** (礼): ritual, propriety, respectful conduct
- **Yi** (义): righteousness, doing what is right regardless of personal gain
- **Zhi** (智): wisdom, discernment grounded in experience and reflection
- **Xin** (信): faithfulness, trustworthiness in word and deed

Virtue, to Confucius, is not about **rigid rules**, but about **harmonious relationships**. It begins in the home, with filial piety *(xiao)*—respect for parents and ancestors—and radiates outward into friendships, governance, and public life.

"To put the world in order, we must first put the nation in order.
To put the nation in order, we must first put the family in order.
To put the family in order, we must first cultivate our personal life."
—*(The Great Learning)*

This is a vision of change that begins not from revolution, but from **inner transformation**.

Li: The Power of Ritual

Confucius placed great emphasis on **ritual**—not as superstition, but as a vessel of meaning. *Li* refers to the structured behaviors that elevate daily life: how one greets a guest, mourns the dead, honors ancestors, or conducts a meeting.

These rituals are not burdensome—they are formative. They teach humility, mindfulness, and respect. They **shape character through action**, creating continuity between the personal and the cosmic.

Through ritual, the ordinary becomes sacred.

Governance as Moral Influence

Confucius believed that a society could only flourish when its leaders were **morally upright**. Law and punishment might control behavior, but only **virtue inspires loyalty**. A good ruler, he taught, is like the north star—stable, consistent, and quietly guiding.

"If the people are led by laws and controlled by punishments, they will try to avoid the punishment and have no sense of shame.
But if they are led by virtue and controlled by the rules of propriety, they will have a sense of shame and will correct themselves."
—*(Analects 2:3)*

Leadership, for Confucius, is not about domination—it is about **serving as a moral example**. The personal and political are inseparable. A corrupt ruler cannot create a just society.

Humanism and Realism

Unlike mystical or metaphysical systems, Confucianism is **grounded in the human world**. Confucius rarely speculated about the afterlife or metaphysical abstractions. When asked about spirits and death, he replied: *"If we do not yet understand life, how can we understand death?"*

This pragmatic humanism has allowed Confucianism to flourish for millennia as a guide for education, governance, and social cohesion—especially in China, Korea, Japan, and Vietnam.

It offers not a path of transcendence, but of **deep presence**—of becoming fully human through attention, intention, and virtue.

The Teacher Endures

Confucius died without seeing his reforms implemented. Yet his legacy has endured like few others. For centuries, his teachings shaped civil service exams, educational philosophy, family structures, and ethical discourse.

He remains, not a prophet or a mystic, but a **model of what it means to teach**—to believe that through word, ritual, and character, the world can be made whole.

His path is not easy, but it is clear: Begin where you are. Cultivate the heart. Respect the sacred in others. And let virtue ripple outward, one life at a time.

Queries for Reflection

- What does it mean to live a virtuous life—not in theory, but in daily practice?
- How do small rituals shape our character, even without words?
- Can moral leadership be restored in a world driven by spectacle and power?
- How might social transformation begin not with revolution, but with personal discipline?

5.6.5
The Analects: Ethics, Ritual, and Social Harmony

The *Analects* of Confucius is not a treatise. It is a conversation. A collection of reflections, exchanges, and remembered sayings, gathered by his disciples after his death. It does not argue or proclaim—it observes. It listens to a teacher navigating the moral challenges of a collapsing world and distilling clarity from experience.

This ancient text, known in Chinese as *Lunyu* (論語), offers not only ethical guidance, but a **portrait of Confucius himself**—humble, persistent, and profoundly devoted to the transformation of both individual and society through the cultivation of **character and conduct**.

A Tapestry of Wisdom

The *Analects* is structured as a series of short passages—some only a sentence long. There is no formal progression, but a thematic rhythm emerges as we move through the text: self-cultivation, filial respect, ethical leadership, restraint, ritual, and the daily practice of becoming fully human.

"The Master said, 'To learn and to practice what is learned time and again, is this not a pleasure?"

—(Analects 1:1)

Learning, for Confucius, is not merely intellectual. It is ethical, relational, and embodied. Knowledge must be lived, not just studied. Virtue must be practiced, not merely praised.

This distinction between knowing and becoming is a central thread in Confucian ethics.

Ethics in Relationship

Confucian morality is never abstract. It is always rooted in **relationship**. A person cannot become virtuous in isolation. Ethics emerges through interaction—with family, friends, elders, subordinates, rulers, and the community as a whole.

"When you see a person of virtue, think to become like them. When you see someone without virtue, examine yourself."
—(Analects 4:17)

This is ethics not as judgment, but as reflection. The actions of others become mirrors in which we refine our own behavior.

At the center of this is the golden thread of **reciprocity**:

"Do not do to others what you would not want done to yourself."
—(Analects 15:23)

This rule, echoed in many spiritual traditions, forms the ethical backbone of Confucian humanism.

Ritual as Social Glue

The *Analects* places strong emphasis on **ritual** (*li*)—not merely religious rites, but the cultivated patterns of daily conduct that hold society together. These include proper greetings, respectful speech, ceremonial acts, and gestures of deference or gratitude.

Confucius believed that when rituals are practiced sincerely, they foster inner refinement. Through outer form, the inner self is trained in humility, patience, and grace.

"Respectfulness, without the rites, becomes laborious bustle.
Caution, without the rites, becomes timidity.
Courage, without the rites, becomes insubordination."
—*(Analects 8:2)*

Ritual, in this view, protects virtue from veering into excess. It keeps human passions and powers **anchored in humility**.

In modern terms, this might be compared to the value of manners—not as superficial gestures, but as the small disciplines that nurture trust, attentiveness, and dignity in everyday life.

The Role of the Junzi

Throughout the *Analects*, Confucius refers often to the **junzi**—the "noble person," sometimes translated as the "gentleman" or "superior man." This is not a person of birth, status, or wealth. The junzi is defined by **inner virtue** and lifelong cultivation.

"The junzi is ashamed to let his words outpace his actions."
—(Analects 14:29)

This figure lives with integrity, honors tradition, practices restraint, and seeks continual improvement. He (or she) is firm but not arrogant, principled but not rigid. The junzi becomes a **moral exemplar** whose life quietly influences others.

In contrast, Confucius warns of the **xiaoren**—the petty person, concerned only with personal gain, flattery, or short-term advantage. The contrast is clear: one builds character, the other masks it.

Governance and Justice

Many passages in the *Analects* address leadership. Confucius believed that **political order must be grounded in moral character**. A good ruler governs less by punishment and more by **example**. Laws may restrain behavior, but only virtue can win hearts.

"If the people are guided by virtue and kept in line by the rites, they will have a sense of shame and will reform themselves."
—(Analects 2:3)

The implication is profound: that society cannot be reformed from the top down or the bottom up alone. It must be reformed from the **inside out**.

True governance begins with self-governance.

The Analects Today

More than two millennia after its compilation, the *Analects* continues to serve as a **handbook for human decency**. It has shaped education, politics, and ethics throughout East Asia—and is increasingly appreciated in the West for its balance of **moral clarity and pragmatic wisdom**.

Its relevance lies not in solving metaphysical questions, but in returning us to essential ones:

- How do we treat others?
- How do we speak and act with consistency?
- How do we build trust, maintain integrity, and live honorably in complex times?

In an era of noise, speed, and spectacle, the *Analects* speaks with a quiet confidence: that real transformation happens not through dramatic gestures, but through **daily attention to the small, the relational, and the real**.

Queries for Reflection

- What does virtue look like in action—not as an ideal, but in daily life?
- How do rituals and customs—past or present—shape our moral awareness?
- Can character still be a foundation for leadership in the modern world?
- What would it mean to live today as a *junzi*—noble in spirit, disciplined in action?

5.6.6
Tao and Li: Two Visions of Wisdom

In the spiritual heritage of ancient China, the teachings of Laozi and Confucius often appear to stand in contrast. One turns to **nature**, the other to **culture**. One dissolves form, the other refines it. One speaks of the **Tao**, the Way of the cosmos; the other of **Li**, the rites and propriety that govern society.

But as with many enduring truths, these paths are not in opposition. They are **complementary visions of wisdom**—offering different lenses through which to perceive the same ultimate aim: a life of harmony, balance, and virtue.

Tao: The Unforced Rhythm of the Real

In Taoism, wisdom means living in accordance with the **Tao**—the nameless, formless origin of all that exists. It is not something to control or define. It is something to **move with**, to trust, to align with.

The Tao does not reward effort or respond to command. It operates like the turning of seasons, the growth of trees, the flow of rivers—quiet, inevitable, vast. It is the underlying **pattern of reality**, and to live wisely is to cease resistance and become part of that flow.

Laozi teaches that the more we seek to control, the more we lose our way. The wise one acts without striving, governs without domination, and speaks without cleverness. The model is not the hero, but the sage—the one who bends, yields, and endures.

Wisdom, in this view, is found in **simplicity, humility, and stillness**. It is not acquired through effort, but uncovered through surrender.

Li: The Patterned Beauty of Human Relationship

In Confucianism, the path to wisdom begins not with retreat into nature, but with engagement in **ritual, relationship, and moral cultivation**. The concept of **Li** refers to a vast field of behavior—ceremony, manners, moral codes, and the subtle rhythms of social life.

Li is what gives shape to virtue. It teaches the body how to bow, the voice how to speak, the mind how to pause before responding. It is the art of becoming **gracefully human** in the midst of family, community, and governance.

For Confucius, Li is not rigid or artificial—it is the **carrier of depth**. When practiced with sincerity, it softens the ego and strengthens character. It allows the self to be shaped by something greater, and offers structure in a world that might otherwise collapse into chaos.

Where the Tao is **cosmic spontaneity**, Li is **cultural intentionality**. Both are ways of aligning with the real—one by yielding, the other by refining.

Stillness and Structure

If Taoism is a path of **emptying**, Confucianism is a path of **filling**. Taoism seeks freedom through detachment. Confucianism seeks harmony through connection. One removes distraction to return to essence. The other engages fully to elevate the everyday.

These approaches reflect not only philosophical divergence, but **temperamental difference**. Some souls are drawn to stillness, others to service. Some find peace in letting go, others in leaning in.

And yet, even Laozi recognizes the value of restraint and humility in leadership. Even Confucius emphasizes the importance of self-cultivation and inner alignment. Their wisdom traditions are not at war—they are **woven through one another**, like yin and yang in the same circle.

A Shared Vision of Harmony

Both Tao and Li aim at **harmony**—within the self, between individuals, and between humanity and the natural world. Neither seeks domination, only alignment. Neither asks for blind obedience, only sincere practice.

One finds peace in the mountain retreat. The other finds it in the ancestral hall. One listens for the wind in the trees. The other listens for the pause between words.

Together, they offer a **holistic path**—reminding us that wisdom is not one thing. It has many textures: flowing and fixed, spontaneous and structured, mystical and moral.

Modern Echoes

In a world divided by extremes—of rigidity and relativism, of control and chaos—the dual vision of Tao and Li offers a middle path.

We can return to nature, but also honor form. We can unlearn distraction, and also relearn civility. We can dissolve the ego, and also train the will.

Wisdom today may mean learning when to act like water and when to show up for the ceremony. When to flow, and when to bow. When to speak freely, and when to pause in reverence.

Each way has its place. And together, they make us whole.

Queries for Reflection

- Do you find yourself more drawn to the spontaneity of Tao or the structure of Li—and why?
- Can letting go and showing up both be acts of wisdom, depending on the context?
- What does harmony look like in your life—not as an idea, but as a daily rhythm?
- How can ancient approaches to wisdom help us navigate modern complexity with grace?

5.6.7
Laozi and Confucius in Comparative and Contemporary Light

Over the centuries, Laozi and Confucius have become more than historical figures. They are symbols—of two archetypal paths that speak not only to the people of ancient China, but to the **human condition itself**. One beckons us to follow the current of the cosmos; the other invites us to build a moral world by our own hands.

Though often framed in contrast, their teachings can also be seen as **interlocking mirrors**, each reflecting different facets of wisdom, wholeness, and harmony. In this section, we step back to view them side by side—not to judge one superior to the other, but to explore how their insights remain profoundly relevant in today's fractured world.

Sage and Scholar: A Comparison of Temperaments

Laozi is the **mystic**: intuitive, enigmatic, elusive. His path is inward, solitary, poetic. He values the hidden over the obvious, the quiet over the loud, the formless over the fixed. His wisdom unfolds in silence and surrender.

Confucius is the **teacher**: ethical, practical, relational. His path is communal, visible, cultivated. He values loyalty, respect, structure, and the transformative power of tradition. His wisdom unfolds in service, structure, and example.

If Laozi is the **river**, winding gently through the landscape, Confucius is the **bridge**, carefully constructed to bring people together. If Laozi dissolves into mystery, Confucius stands firmly in history.

And yet, both are responding to the same crisis: a society out of balance, a people unmoored, a world in need of **re-orientation**.

Their divergence lies in method, not purpose.

East and West: Parallels and Refractions

Across cultures, Laozi and Confucius find echoes in other great traditions:

- **Laozi** can be compared to **Heraclitus**, the Greek philosopher who spoke of the unity of opposites and the ever-changing flow of life. His cryptic aphorisms resonate with Taoist paradox.
- He also aligns with **mystics** across traditions—**Meister Eckhart** in Christianity, **Rumi** in Islam, **Shankara** in Hinduism—each exploring the dissolution of the ego and union with the ineffable.
- **Confucius**, by contrast, resonates with thinkers like **Socrates** and **Marcus Aurelius**, who emphasized virtue, discipline, civic responsibility, and the shaping of ethical character through reason and dialogue.
- His vision also parallels the **Jewish prophetic tradition**, where the moral structure of the world is built through justice, remembrance, and righteous action.

These parallels suggest something important: that **truth has many expressions**, and that Laozi and Confucius—while rooted in their cultural soil—are part of a broader human search for meaning, coherence, and flourishing.

In Dialogue with Modernity

In the modern world, both paths offer antidotes to imbalance.

Laozi speaks powerfully to a culture of overdrive, burnout, and hypercontrol. He reminds us that much of life unfolds best when we **let go**—when we learn to yield, to listen, to trust what is natural, cyclical, and quiet.

In moments of anxiety or disconnection, Taoism offers a return to the present: the breath, the breeze, the effortless unfolding of life when we stop pushing.

Confucius, on the other hand, addresses a different kind of crisis—the erosion of **civility, responsibility, and intergenerational respect**. In an age of moral relativism and social fragmentation, his call to ethical development, thoughtful ritual, and noble character provides a grounded path forward.

He reminds us that integrity is not obsolete. That discipline and grace matter. That **how we live together** is a sacred project.

Integration, Not Opposition

Rather than seeing Laozi and Confucius as opposites to choose between, we can see them as **complementary forces**—like yin and yang, form and emptiness, action and stillness.

Their dialogue continues not only in China but across the globe, wherever people seek balance between the **outer structures of society** and the **inner rhythm of being**.

To live well in this world may require both:

- The clarity and accountability of Confucius
- The surrender and spontaneity of Laozi
- The ability to build, and the wisdom to yield
- The capacity to bow at the altar, and the ability to flow with the river

Together, they form a bridge between **ethics and mysticism**, between doing and being—each offering what the other lacks, each reminding us of a different way to be whole.

Queries for Reflection

- How do Laozi and Confucius speak to different aspects of your own journey?
- Are there times when one path—structure or flow—needs to guide the other?

- What does a balanced life look like when informed by both ethical discipline and cosmic surrender?
- In a polarized world, can we learn to hold opposites in creative tension, as these two sages did?

5.6.8
Eastward Inquiry: Living Between Flow and Form

In the teachings of Laozi and Confucius, we find not just two philosophies, but two **orientations to life**—two ways of walking the path, two voices that continue to speak to the modern soul. As we complete our exploration of these traditions, we are left with an open invitation: to live **between flow and form**, not as a contradiction, but as a dance.

Walking the Middle of the Way

It is tempting to lean entirely toward one worldview or the other.

Laozi calls us to **let go**, to yield, to stop striving. His wisdom soothes those exhausted by systems, perfectionism, and pressure. To embrace Taoism is to remember that much of life takes care of itself when we stop forcing outcomes.

Confucius, on the other hand, urges us to **step up**—to take responsibility, to refine our behavior, to lead by moral example. His wisdom grounds us when we feel aimless or fragmented. It reminds us that freedom without responsibility can become emptiness.

But perhaps the deepest wisdom lies not in choosing between them, but in **holding both**.

We need **ritual** and **release**, **discipline** and **emptiness**, **character** and **compassion**, **precision** and **presence**.

To live between flow and form is to **adapt**—to know when to act with clarity and when to wait with trust. It is to navigate life with both ethical grounding and spiritual openness.

Contemporary Applications

In leadership, we can apply Confucian values of accountability and respect, while also drawing on Taoist humility and flexibility.

In relationships, we can honor the rituals of connection and care (Li), while allowing room for silence, spontaneity, and authenticity (Tao).

In personal development, we might embrace a Confucian commitment to growth—not through shame, but through steady refinement—while also remembering that we are already part of a larger rhythm, whole even in our becoming.

In our spiritual lives, we might pray with form and reverence, but also sit quietly and listen—to wind, to breath, to Being itself.

This is not compromise. It is **integration**.

The Spirit of Inquiry

The purpose of this chapter—and this series—is not to tell you what to believe. It is to **invite reflection**. To offer glimpses of enduring wisdom that still pulse with relevance. To trace lines between ancient questions and modern concerns.

In the case of Laozi and Confucius, the inquiry continues:

- What does it mean to live well?
- How do we align with something larger than ourselves?
- What makes a society whole—and what makes a life worth living?

The answers are not fixed. They evolve, as we do.

But what these sages offer is a starting point. A deep well of insight. A mirror in which we may see ourselves—not to flatter or shame, but to **remember**.

Remember that stillness is not weakness. That courtesy is not empty. That nature has lessons for us. That tradition can be a guide. That balance is not found in certainty, but in openness.

A Closing Reflection

Imagine a life where you rise each day with purpose but without tension.
Where your words are few, but well-chosen.
Where you greet your ancestors and greet the dawn.
Where you act with clarity, then let go.
Where you bow with sincerity, and then laugh like water over stone.

This is the invitation of Eastward Inquiry.
Not to become someone else, but to **remember what is already within you**.
To find your way—through the stream, through the ceremony, through the stillness that holds them both.

Final Queries for 5.6

- What balance of flow and form does your life currently reflect?
- Where have you over-structured, and where have you over-drifted?
- Can you hold reverence without rigidity, and spontaneity without chaos?
- What small practices—of stillness or service—might bring you back into rhythm?

7.

GURU NANAK

The One Light Through Ten Forms

-ZEPHYRO SOLON

CONTENTS

5.7.1
India in Transition: Spiritual Landscape of the 15th Century

When Guru Nanak was born in 1469, India stood at a crossroads of cultural, spiritual, and political transformation. The subcontinent was a mosaic of kingdoms, traditions, and tensions. It was a place of **deep devotion** and **deep division**, where ancient scriptures echoed through temples and forests, even as ordinary people struggled beneath the weight of ritualism, caste, and conflict.

This was a time of both **ferment and fragmentation**—a fertile ground for reformers, saints, and mystics who sought to return the people to the essence of the sacred.

A Landscape of Many Paths

Religiously, the Indian landscape was rich and varied. **Hinduism**, with its many deities, philosophies, and texts, was deeply woven into the fabric of daily life. From the Vedas to the Upanishads, from temple worship to household rituals, its influence was vast—but not without internal critique.

The **bhakti movement** had already begun to ripple across India, offering a heartfelt, personal connection to the Divine that transcended caste and priestly control. Saints like **Kabir**, **Mirabai**, and **Tulsidas** were singing songs of love to God—without intermediaries, without boundaries.

At the same time, **Islam** had established itself in northern India through centuries of trade, migration, and conquest. The **Delhi Sultanate** and the rise of **Mughal rule** brought Islamic art, architecture, poetry, and theology into contact—and sometimes collision—with Indian traditions.

Sufism, the mystical heart of Islam, also flourished in the region. Sufi saints like **Nizamuddin Auliya** and **Moinuddin Chishti** offered a message of divine love and inner surrender, often attracting Hindu and Muslim followers alike.

In this pluralistic, often polarized world, many ordinary people were drawn to **direct experience**—to a God who could be felt, heard, sung to—not merely described in books or mediated by hierarchy.

Social Tensions and the Caste Divide

While spiritual light abounded, so too did social darkness. The **caste system** remained deeply entrenched, enforcing a strict hierarchy of birth and purity that shaped everything from one's profession to one's place of worship. Many were excluded from education, dignity, and participation in spiritual life.

Women faced severe restrictions, and interfaith tensions were growing between Hindu and Muslim communities. Religious identity had become a source of both pride and division.

Amid this complexity, spiritual truth seemed to have been **buried under form**. People were hungry for something simpler, clearer—something that spoke directly to the soul.

It is into this world—vibrant, wounded, searching—that Guru Nanak was born.

The Call for a New Synthesis

Nanak did not arrive to start a religion. He came to **reawaken unity**. His message did not reject Hinduism or Islam, but **transcended both**, calling people back to the Oneness that lies beneath all names and forms.

He spoke of **Ek Onkar**—One Reality, endlessly expressed. He challenged caste, privilege, and spiritual pride. He walked across India, Persia, Tibet, and Arabia not to conquer, but to listen, to sing, to share.

In Nanak, we see not a reaction, but a revelation—emerging not from opposition, but from **deep seeing**.

His life and teachings reflect the **soul of his time**, and yet they rise beyond it—offering a vision of unity, simplicity, and service that still speaks across boundaries of faith, language, and time.

Queries for Reflection

- What happens when spiritual truths become entangled with social division?
- Can devotion and justice be part of the same sacred path?
- What cultural tensions today resemble the ones Nanak inherited—and how might we respond?
- What does it mean to speak a spiritual language that transcends religious boundaries without denying them?

5.7.2
The Birth and Early Life of Nanak

Guru Nanak was born in 1469 in the village of **Talwandi**, now known as **Nankana Sahib**, in present-day Pakistan. From the beginning, his life radiated with a sense of quiet mystery and gentle power. Though his family was of the Hindu merchant caste (*Khatri*), Nanak would grow into a soul who belonged to no division—and to all people at once.

The world into which he was born was one of **expectation and confinement**: caste roles were fixed, rituals were rigid, and spirituality often came wrapped in complex dogma. But from a young age, Nanak seemed **attuned to something deeper**—a presence beneath form, a unity beneath difference.

Signs of a Seeker

Stories from his childhood, preserved in the Sikh tradition, speak of a child **uninterested in wealth or status**, one who preferred silence and reflection over business or ritual. When given sacred threads during a Hindu coming-of-age ceremony, Nanak is said to have refused, asking instead why such threads should matter if one's heart and actions were not pure.

"Let mercy be your cotton, contentment your thread,
Truth the knot, and virtue the twist."
—(Guru Granth Sahib, p. 471)

From early on, Nanak was not rebellious—he was **awake**. He was not cynical—he was **sincere**. His questions were not challenges to authority for their own sake, but invitations to return to what was **real**.

He showed compassion to all creatures, treated servants and sages alike, and displayed an uncanny **equanimity**—a calm that did not withdraw from the world, but saw through it.

A Glimpse of the Beyond

According to tradition, around the age of thirty, Nanak underwent a **profound spiritual experience** that would alter the course of his life—and the lives of millions. One morning, while bathing in a nearby river, Nanak disappeared. His companions searched, fearing he had drowned. But three days later, he emerged—silent, radiant, changed.

His first words upon returning were simple and monumental:

"There is no Hindu, there is no Muslim."

It was not a denial of religious identity, but a **revelation of deeper unity**. He had seen beyond forms to the **One Light** that shines through all paths. From that moment, his life was dedicated not to founding a sect, but to pointing humanity back to **truth beyond division**.

He spoke of **Ek Onkar**—the One Reality—and called people not to conversion, but to **awakening**.

Family, Tensions, and Early Teachings

Though Nanak's mystical insight set him apart, he did not retreat from the world. He married **Mata Sulakhni** and had two sons. But even as he fulfilled the role of householder, his heart remained devoted to the **inner call**.

His brother-in-law, skeptical of his spiritual leanings, helped secure him a job as a storekeeper for a local Muslim governor. Nanak served honestly—but stories tell of him giving freely to the poor, unable to see God's creation divided into rich and poor, worthy and unworthy.

In this, we glimpse the seeds of **Seva**—selfless service—that would become core to the Sikh path. Even in his early years, Nanak lived as if God was **not in a temple, but in every face**.

Those around him often misunderstood his stillness. Yet his silence was not emptiness. It was **listening**—to the eternal song beneath the noise of the world.

Toward a Life of Message

By his early thirties, Nanak was preparing—not with announcements or dogma, but with depth. His heart was full of song, his words were beginning to flow in verse, and the path ahead was being revealed.

He had seen something—beyond division, beyond ritual, beyond fear. Now he was ready to **walk and share**. Not to escape the world, but to awaken it.

The one who had once bathed in a river would now cross **rivers of culture, language, and tradition**—carrying not a religion, but a resonance.

Queries for Reflection

- How does early sensitivity to injustice or ritual emptiness shape a spiritual path?
- What might it mean to see no Hindu, no Muslim—but only humanity and Light?
- Can family life and spiritual purpose coexist—not in tension, but in mutual support?
- Where in your own story has silence, retreat, or mystery prepared you to speak?

5.7.3
The Call to Wander: Nanak's Journeys and Teachings

After his transformative awakening at the river, Guru Nanak did not claim a throne, start a religion, or seek followers. Instead, he **walked**. Leaving behind his worldly duties, family, and home—not in rejection, but in response to the **voice within**—he began a journey that would span decades, thousands of miles, and countless conversations.

These travels, called the **Udasis**, were not political missions or pilgrimages. They were sacred wanderings—rooted in the conviction that the Divine is not confined to any one place, creed, or language. God is present wherever truth is spoken, justice is honored, and the heart is open.

Nanak's life became **a moving sanctuary**, and his presence a quiet revolution.

The Spirit of the Udasis

Tradition holds that Nanak undertook at least four major journeys, traveling throughout the Indian subcontinent and beyond. He visited:

- Hindu temples and sacred riverbanks in India
- Sufi shrines and Islamic holy cities, including Mecca and Medina
- Tibetan Buddhist centers and Himalayan villages
- Jain monasteries, marketplaces, and courtrooms

Everywhere he went, he **listened first**. He asked questions, shared food, observed customs, and only then began to speak—often in the form of **poetic song** (*shabad*), drawing from the language of the people and the rhythms of the Divine.

He traveled not with wealth or power, but with a companion—**Mardana**, a Muslim musician and dear friend. Mardana would play the **rabab**, a stringed instrument, while Nanak sang. Together, they wove a message of **unity through harmony**—across religions, castes, and borders.

In Nanak's company, distinctions began to blur. The sacred was no longer locked in texts or temples. It became **alive in encounter**.

Confronting Corruption with Grace

Nanak's teachings were simple but piercing. He spoke against **empty rituals**, caste oppression, religious hypocrisy, and material greed. But he did not shout or shame. He used **metaphor, humor, and humility**—holding up a mirror with gentleness and power.

To a wealthy priest obsessed with appearances, he might say:

"Without inner cleansing, wearing sacred threads is like putting perfume on a corpse."

To a soldier burdened by violence, he would sing of courage that defends the weak, not power that dominates.

To devotees lost in rules, he would ask, "Can your fast feed the hungry? Can your prayer heal your neighbor?"

His message was not destructive. It was **clarifying**. He did not seek to tear down faith, but to **restore its flame**.

Ek Onkar: Unity as Foundation

At the heart of Nanak's message was the profound affirmation of **Ek Onkar**: One Reality. One Source. One Light shining through all forms.

This was not a bland universalism. It was **a lived realization**—that behind the many names and forms of God, behind the divisions of religion and culture, there is only **Oneness**.

"There is one Creator of all creation, whose name is Truth.
The Timeless One, without fear or hatred, beyond birth and death, self-existent,

realized by grace."
—(Japji Sahib, opening lines)

For Nanak, God was not confined to the heavens or the scriptures, but **immanent and personal**—nearer than breath, and accessible to all.

Prayer, then, was not ritual performance, but remembrance. Worship was not ceremony, but **alignment**—a way of walking through the world with humility, love, and awareness.

Everyday Miracles

Though many miracles are attributed to Nanak—appearing in multiple places at once, turning the direction of Mecca by his mere presence, calming storms—they were never the point. What mattered was **transformation**—not of nature, but of the heart.

Wherever he went, people were changed. A merchant would give up hoarding. A priest would begin to serve the poor. A warrior would lay down his pride. Not because Nanak forced them to, but because something in his **presence** called forth their deeper self.

He did not build temples, but he **planted seeds**—of dignity, simplicity, devotion, and fearlessness.

Not Conversion, but Awakening

Nanak never asked anyone to abandon their tradition. He never claimed exclusivity or demanded allegiance. He simply pointed to the **One behind the many**. To the truth that all names—Allah, Ram, Waheguru—can open the same door when spoken with love.

His journey was a reminder that **the real pilgrimage is inward**—and that the sacred can be encountered in a Muslim, a Hindu, a stranger, a servant, a sunrise.

To follow Nanak was not to join a group. It was to join a **way of seeing**.

And to sing, with each breath: There is only One.

Queries for Reflection

- Can sacred truth be carried not in doctrine, but in song, presence, and humility?
- What does it mean to travel—not to escape, but to bear witness and connect?
- How might we speak truth to power without aggression—through clarity and compassion?
- Are we willing to walk beyond the borders of religion to rediscover the unity beneath?

5.7.4
Ek Onkar: The One Reality and the Essence of Sikh Thought

At the very beginning of the Guru Granth Sahib—the central scripture of Sikhism—appear two words of profound simplicity and cosmic scope: **Ek Onkar**.

It means, literally, **"One Reality"** or **"One Creator."** Not just one god among many, but **the singular, formless, timeless essence** that pervades all creation. These words, spoken by Guru Nanak after his awakening, are more than a theological claim—they are a **vibrational doorway** into a worldview of unity, devotion, and spiritual clarity.

Everything that followed—his teachings, his travels, the formation of the Sikh community—emerged from this realization: That behind every name, every form, every boundary, there is **One Light**.

The Mool Mantar: A Map in a Mantra

Ek Onkar is the opening of what is known as the **Mool Mantar**, or "Root Mantra"—a concise yet vast poetic statement that outlines the essence of Sikh understanding. In its full form, it reads:

"Ek Onkar
Sat Naam
Karta Purakh
Nirbhau
Nirvair
Akaal Moorat
Ajooni

Saibhang
Gur Prasaad"

Each phrase opens a new dimension:

- **Ek Onkar** – One Reality, One Creator
- **Sat Naam** – True is the Name, or Identity
- **Karta Purakh** – The Creative, Active Being
- **Nirbhau** – Without fear
- **Nirvair** – Without enmity
- **Akaal Moorat** – Timeless form
- **Ajooni** – Unborn, beyond birth
- **Saibhang** – Self-existent
- **Gur Prasaad** – Realized by the Guru's grace

This is not a creed to memorize, but a **path to walk**. It reveals not only what the Divine is—but what we are invited to become. To live in alignment with Ek Onkar is to **see clearly**, act justly, love freely, and dissolve the false divisions between "self" and "other."

Oneness Beyond Division

In a world fractured by caste, creed, gender, and power, Nanak's declaration of Ek Onkar was both **spiritual and revolutionary**.

It meant:

- There is no superior religion.
- No higher or lower birth.
- No separation between sacred and ordinary.
- No justification for hatred, because all are expressions of the same Light.

This Oneness was not theoretical. It was lived:

- In **shared meals** (*langar*), where all sat side by side regardless of status.
- In **service to others**, without expectation of reward.
- In **daily remembrance**, where work and worship were integrated.

The Sikh path would grow around this center—not as a rejection of Hinduism or Islam, but as a new **stream of clarity**, fed by many rivers yet flowing in its own rhythm.

Name as Presence

One of the most distinctive features of Sikh spirituality is the practice of **Naam Simran**—the remembrance of the Name (*Naam*), understood not merely as a word, but as the **vibrational presence** of the Divine.

To repeat the Name—Waheguru, Sat Naam, or simply the breath itself—is to return the mind to its Source. It is not magic, but **attunement**.

Where some traditions rely on theology or ritual, Sikh practice emphasizes **awareness**—a continual return to the One, in thought, word, and action. The goal is not withdrawal from life, but **illumined participation** in it.

"Why do you go to the forest in search of God?
God lives in all and abides with you too.
Like fragrance in the flower,
Like reflection in the mirror,
God is revealed in all things."
—(Guru Nanak)

Oneness as Ethics

The affirmation of One Reality also grounds a deep **ethic of equality and justice**. If all are expressions of the One, then to harm another is to harm the Divine. If all beings carry the same Light, then compassion, service, and courage are not optional—they are natural outcomes of **clear seeing**.

This ethic would guide not only Guru Nanak, but the Gurus who followed him—who established communal kitchens, armed the oppressed, and lived lives of **fearless devotion** in the face of injustice.

Ek Onkar is not only a spiritual realization—it is a **call to action**. A vision that unites contemplation and compassion, silence and service, prayer and participation.

A Truth for Every Time

Today, in a world once again marked by division and anxiety, Ek Onkar remains a **timeless mantra**. It invites us to release the illusion of separateness, to bow to the same Light in every face, and to align our actions with the rhythm of unity.

It is both anchor and horizon: a place to return to, and a path that never ends.

Queries for Reflection

- What would it mean to live as if all beings truly came from One Source?
- Can repeating the Name become a path not of escape, but of deep presence?
- How might our ethics change if we saw no separation between spirit and society?
- What does "Oneness" ask of you—not as a belief, but as a practice?

5.7.5
Householder, Mystic, Revolutionary

Guru Nanak's life defied categories. He was not a monk, nor a renunciate. He did not retreat from the world, nor immerse himself in its illusions. He lived as a householder, walked as a mystic, and acted as a quiet revolutionary—embodying a path that wove together the spiritual and the everyday.

In doing so, he laid the foundation for a tradition that is not merely contemplative, but **integrated**—a path where the sacred is not something we visit, but something we **live**.

The Sacred Within the Ordinary

Unlike many spiritual figures of his time, Nanak did not reject family life. He married **Mata Sulakhni**, had two sons, and engaged in the rhythms of domestic responsibility. This was not a compromise—it was a **principle**. He taught that liberation did not require withdrawal, but **awareness within activity**.

To live in the world while remaining inwardly free—that was the ideal.

In this sense, Nanak offered a **new model of spiritual life**: one where the mystic is not hidden in caves, but raising children, tilling soil, serving meals, and remembering the One in all of it.

This became a cornerstone of Sikh thought: that one can live fully in the world—engaged, productive, loving—and still remain **anchored in the Divine**.

Work was not separate from worship. It was a **form of it**.

The Mystic Path: Inner Fire and Divine Union

Though Nanak lived simply, his inner life burned with intensity. His poetry is filled with longing for the Divine Beloved, awe before the majesty of the universe, and the ecstasy of union.

"When the mind is dyed in the color of the Name,
It is absorbed in the Lord as water merges with water."
—(Guru Granth Sahib, p. 349)

His teachings reflect the **nondual awareness** found in many mystical traditions: that the soul and the Divine are not separate, and that liberation lies not in reaching some distant heaven, but in **recognizing the Light within**.

He often described the human being as a musical instrument—capable of great resonance when tuned to the Divine frequency. Prayer, song, and service were ways of tuning this instrument, aligning it with the eternal current.

Nanak's mysticism was not esoteric. It was **accessible**. It invited everyone, regardless of caste or gender, to remember their divine origin and walk the path of love.

The Quiet Revolutionary

And yet, Nanak's vision was not only personal. It had profound **social and political implications**. To affirm the equality of all beings was to challenge entrenched hierarchies. To reject caste was to confront the foundations of Hindu orthodoxy. To proclaim that God is neither Hindu nor Muslim was to **stand between worlds**—and call them to account.

His teachings challenged:

- The **ritualism** of religious elites
- The **oppression** of lower castes
- The **patriarchy** that silenced women
- The **dogma** that divided people in the name of God

But he did not preach rebellion. He lived an alternative—peacefully, clearly, unwaveringly.

His emphasis on **Naam (remembrance), Kirat (honest work), and Vand Chakna (sharing with others)** became spiritual principles and **ethical blueprints**. They were not abstract ideals, but **daily disciplines** that transformed the individual and, through them, the community.

A New Kind of Spiritual Authority

Nanak never claimed divinity. He did not perform for attention. He did not seek followers, yet they came—drawn not to charisma, but to **authenticity**.

He taught through poetry, presence, and example. His authority came not from hierarchy, but from **alignment**—from the sense that he was living what he spoke, embodying what he sang.

This kind of authority is rare. It is the kind that doesn't dominate but **awakens**.

Legacy in Motion

By the time of his death in 1539, Guru Nanak had planted a vision of **divine equality and spiritual clarity** that would be carried forward by nine successive Gurus—each expanding, protecting, and deepening the path.

But the essence remained his:

- The integration of spirit and service
- The celebration of God within the everyday
- The courage to speak truth without violence
- And the invitation to see the One in all

He left no empire. He left no scripture penned by his own hand. He left **a vibration**, a way of walking.

And it continues—wherever someone bows to the One, works with honesty, serves without pride, and sings the Name with love.

Queries for Reflection

- Can mysticism thrive not in retreat, but in the heart of family and work?
- What does it mean to resist injustice not through protest alone, but through living differently?
- How might our lives become more integrated—where prayer and action, silence and labor, are one?
- What would it mean to live with both fire and gentleness, like Nanak?

5.7.6
The Guru Granth Sahib and the Living Word

When Guru Nanak's light passed from this world, it did not disappear. It became a lineage—not of blood, but of **spiritual transmission**. Nine Gurus followed him, each carrying forward the vision, deepening its message, and guiding the growing Sikh community through times of expansion and persecution.

But unlike many spiritual traditions centered around a single figure or book, Sikhism would evolve into something unique: a path where the **scripture itself became the Guru**—the eternal teacher.

This scripture, the **Guru Granth Sahib**, is not merely a sacred text. It is revered as the **living embodiment of divine wisdom**, a vibrational presence, a Guru in its own right. For Sikhs, to sit before it is to be in the presence of truth. To hear it is to be guided. To sing it is to merge with the Divine.

A Scripture Like No Other

The Guru Granth Sahib is a compilation of over 1,400 pages of poetic hymns (*shabads*), composed not only by the Sikh Gurus, but also by saints and mystics from diverse traditions—**Hindu bhaktas, Muslim Sufis, low-caste poets, and more**.

This inclusion was radical. It affirmed what Nanak taught from the beginning: that **truth is not owned by any one tradition**, and that the Divine speaks through sincerity, not lineage.

Some of the contributors include:

- **Kabir**, a weaver and poet-saint known for his fierce critiques of hypocrisy

- **Ravidas**, a leatherworker from a marginalized caste, whose devotion and clarity shine through every line
- **Sheikh Farid**, a Sufi mystic whose verses blend asceticism and sweetness

The message is consistent: love the One, serve all, and realize the Divine **within and around you**.

The scripture is written in **Gurmukhi script**, using a blend of languages—Punjabi, Persian, Sanskrit, and Braj—reflecting the linguistic diversity of its time and the **universality of its message**.

The Word as Vibration

In Sikhism, the **Shabad (Word)** is not static text. It is a **living vibration**—a sound current that carries the presence of God. To read the Guru Granth Sahib is not merely to understand; it is to **resonate**, to allow the words to **re-tune the mind and heart**.

This is why music is central to Sikh practice. The scripture is organized by **rāgas**—musical modes—designed to evoke specific moods and spiritual states. Singing or listening to these hymns becomes a form of **Naam Simran**, remembrance of the Divine through sound.

It is not intellectual analysis, but **embodied devotion**. A heart-opening. A return.

"The Word is the Guru, and the Guru is the Word.
Within the Word is the nectar; the Word brings union with the Lord."
—(Guru Granth Sahib, p. 982)

The Guru Beyond Person

In 1708, the tenth Guru, **Guru Gobind Singh**, formally ended the line of human Gurus and installed the Guru Granth Sahib as the **eternal Guru**. From then on, no individual would hold exclusive spiritual authority. The community—guided by the Word—would walk forward **together**.

This was a revolutionary act. It decentralized power, affirmed equality, and placed the focus not on charisma, but on **shared devotion and discernment**.

The Guru became:

- The **Granth**: the scripture
- The **Panth**: the community of seekers
- The **Presence**: the Light accessed through the Word and lived through action

In a world of religious hierarchy and personality cults, this was—and remains—profoundly liberating.

The Granth in Daily Life

In gurdwaras (Sikh temples) around the world, the Guru Granth Sahib is placed with reverence upon a platform, covered with beautiful cloths, and read aloud each day. Attendees listen quietly, not as spectators, but as participants in **a sacred dialogue**.

Weddings, births, deaths, and major life transitions are all centered around readings from the Granth, often chosen randomly in a practice called **hukam**—seeking guidance from wherever the scripture opens.

But the real teaching of the Granth is not bound to ceremony. It is found in the **practice of its wisdom**:

- In working honestly
- Sharing what you earn
- Serving others with humility
- And remembering the Divine through word, work, and love

A Living Guide for All

The Guru Granth Sahib is not reserved for Sikhs alone. Its wisdom speaks to **any seeker** longing to live with clarity, compassion, and presence. It is a scripture that **includes other voices**, and in doing so, opens a space for **each reader to be transformed**.

It does not ask you to adopt a label. It asks you to return to the Name.
Not in abstraction. But in the middle of life.
In your work. In your kitchen. In your silence. In your song.

Queries for Reflection

- What does it mean to treat a text as a living presence, not just printed words?
- Can music become a path to transformation—not just entertainment, but communion?
- How might we listen to sacred texts—not to be right, but to be realigned?
- What wisdoms might we miss when we seek truth only in our own tradition?

5.7.7
The Legacy of the Ten Gurus and the Spirit of Seva

Though Guru Nanak was the first to speak the truth of Ek Onkar, he did not carry that vision alone. After him came **nine successive Gurus**, each entrusted not with altering the message, but with **embodying and expanding it**—guiding the Sikh community through unfolding challenges, new horizons, and deeper realizations.

Together, the Ten Gurus created a **living arc of transmission**, a spiritual lineage rooted in humility, courage, and service. They did not reign from thrones. They walked among the people. They served, sang, resisted injustice, and built the foundations of one of the world's most distinctive and resilient spiritual paths.

Ten Gurus, One Light

Each Guru brought something unique, yet all carried the same Light—the same commitment to the One Reality, to justice, to devotion, and to the dignity of all beings.

A few highlights:

- **Guru Angad Dev**, the second Guru, preserved Nanak's teachings and developed the **Gurmukhi script**, enabling widespread literacy and access to sacred texts.
- **Guru Amar Das**, the third Guru, abolished caste distinctions in the Sikh community and institutionalized **langar**—the communal kitchen where all people, regardless of background, eat together.

- **Guru Ram Das**, the fourth, composed beautiful devotional poetry and founded the city of **Amritsar**, which would become a spiritual center for Sikhs.
- **Guru Arjan Dev**, the fifth, compiled the **Adi Granth**, the precursor to the Guru Granth Sahib. He was also the first Sikh martyr, executed by Mughal authorities for his refusal to abandon the faith.
- **Guru Hargobind**, the sixth, introduced the concept of **miri and piri**—the balance of spiritual and temporal power. He wore two swords to symbolize this integration, teaching that a saint may also need to be a soldier.
- **Guru Har Rai** and **Guru Har Krishan**, the seventh and eighth Gurus, emphasized healing, compassion, and care during times of political turmoil and disease.
- **Guru Tegh Bahadur**, the ninth, gave his life to defend the religious freedom of others—including Hindus—when faced with forced conversions by the Mughal empire. His martyrdom is remembered as an act of **selfless courage** in the name of universal dignity.
- **Guru Gobind Singh**, the tenth and final human Guru, transformed the Sikh community into the **Khalsa**, a collective of spiritually disciplined, courageous warriors of peace. He affirmed that after him, **the Guru would live in the Word**—in the Guru Granth Sahib.

Seva: The Practice of Sacred Service

Across all ten Gurus, one theme shines consistently: **Seva**—selfless service.

In Sikh tradition, Seva is not charity. It is not optional. It is **worship in action**—a way of dissolving ego, honoring the One in all, and transforming both the giver and the world.

Seva may take many forms:

- Cooking and serving in a gurdwara's langar
- Cleaning the shoes of visitors
- Providing aid to those in need, regardless of identity

- Standing against injustice
- Offering one's work, presence, and time without expectation of reward

This practice is deeply woven into Sikh life. From small villages in Punjab to bustling global cities, Sikhs are known for showing up wherever help is needed—feeding protesters, delivering aid in disaster zones, offering water at festivals, and standing as guardians of the vulnerable.

Why? Because **God dwells in everyone**, and to serve another is to serve the Divine.

"The hands that serve are holier than the lips that pray."
—(Guru Nanak)

Seva is not dramatic. It is not self-congratulatory. It is **quiet strength**, born from inner alignment. It does not ask for applause—it asks, *Who needs me now?*

The Warrior-Sage Ideal

With the creation of the **Khalsa** in 1699, Guru Gobind Singh solidified a powerful integration: the spiritual seeker and the defender of truth. He taught that devotion must be **fearless**, and that love of God must include **protection of the oppressed**.

The Khalsa were given five symbols of commitment (the **Five Ks**) and a new last name—**Singh** for men (lion), **Kaur** for women (princess)—abolishing caste identifiers and affirming equality.

Yet even in this martial form, humility remained central. The sword was to be **wielded with compassion**, and only in defense of the innocent.

This warrior-sage model reflects the Sikh ideal: a person who **remembers the Name, walks with grace, serves without ego, and stands with courage**.

A Living Legacy

The Gurus are not distant figures. They are **living inspirations**, accessible through the Granth, through community, and through the disciplines of remembrance, service, and integrity.

Their legacy is not frozen in time. It lives wherever someone:

- Works honestly
- Shares freely
- Bows to no false authority
- Refuses to be silent in the face of cruelty
- Washes dishes with devotion
- Speaks truth with gentleness
- Serves a stranger as an embodiment of the One

Queries for Reflection

- What would it mean to make your life a form of service—not out of duty, but out of devotion?
- How do humility and courage work together in your spiritual path?
- Can everyday acts—cooking, cleaning, helping—be transformed into sacred offerings?
- What parts of you are being called to live more like a servant—and more like a guardian?

5.7.8
Walking with Nanak: Unity, Courage, and the Eternal Sound

Guru Nanak did not leave behind a dogma. He left a **direction**—a way of walking. His path was not one of escape from the world, but of **deep engagement** with it. His footsteps echoed with unity, courage, and remembrance—qualities that continue to guide millions, not only in temples and scriptures, but in kitchens, protests, quiet hearts, and daily acts of kindness.

To walk with Nanak is not to follow a fixed set of rules. It is to awaken to a **certain way of seeing**—one that perceives the Divine not just in the sky, but in the face of every stranger. Not just in prayer, but in labor. Not just in solitude, but in service.

It is to see **no Hindu, no Muslim**, no higher or lower—but only One Light, refracted through many forms.

The Sound of the One

Nanak taught that the world is sustained by **Naam**—the eternal sound, the Divine vibration, the resonance of truth that permeates all things.

To attune oneself to Naam is not to retreat from life. It is to live **from the center of it**—with awareness, humility, and love.

"When the mind is attuned to the Lord,
It becomes like the lotus untouched by the water."
—(Guru Granth Sahib, p. 938)

The practice of **Naam Simran**—repeating and remembering the Name—is not escapism. It is a kind of spiritual tuning. In the midst of the noise and demands of modern life, it brings us back to the **eternal sound** that grounds all action.

Unity in the Midst of Division

Nanak's message remains **radical** today, not because it is exclusive, but because it is **inclusive without dilution**. He did not erase difference, but illuminated the unity beneath it.

In a world still torn by religious conflict, racial hierarchy, economic disparity, and spiritual exhaustion, his path offers clarity:

- Remember the One.
- Serve the many.
- Speak truth with gentleness.
- Stand for the oppressed.
- Live simply.
- Work honestly.
- Sing often.
- And see the Divine in everyone.

A Warrior's Grace

To follow Nanak is to live with **strength and softness**—to act without fear, but never without compassion. The courage he exemplified was not about domination. It was about **resistance through love**.

Sikhs have carried this legacy into every age—standing against tyranny, feeding the hungry, opening their doors to all, and walking through the world not with swords of pride, but with the **spirit of seva** and the clarity of Naam.

This is not passive spirituality. It is active grace.

The Path Ahead

You do not need to wear a turban to walk with Nanak. You do not need to be born in Punjab or raised in any tradition.

You need only to **listen inwardly**, to love without limit, and to serve without seeking reward.

You need only to let your work become worship, your speech become song, your actions become offerings, and your breath become remembrance.

Nanak walked across empires without conquering. He walked across religions without erasing. He walked across time—and he is still walking.

The question is: **Will we walk with him?**

Final Queries for 5.7

- What would it mean to walk through life as Nanak did—with presence, courage, and compassion?
- Can you live in a way that dissolves boundaries while still honoring form?
- How might you bring the remembrance of the One into the smallest acts of your day?
- What song is your soul being called to sing—and who will hear it through your life?

8.

THE SEVEN SAGES OF INDIA

Keepers of Cosmic Memory, Guardians of the Vedic Flame

-ZEPHYRO SOLON

CONTENTS

5.8.1
Myth, Memory, and Cosmic Lineage: Who Were the Saptarishi?

Long before the rise of recorded history, before the teachings of the Buddha or the songs of the Vedas were formally written, the Indian tradition speaks of **seven luminous beings**—the **Saptarishi**, or "Seven Sages"—who emerged not from the earth but from **cosmic order itself**.

They are not simply ancient wise men. They are mythic ancestors, archetypal seers, and spiritual bridges between **heaven and humanity**. According to Hindu cosmology, they were born of the mind of Brahma, the Creator, and entrusted with **preserving dharma**, transmitting divine knowledge, and guiding the early civilizations of Earth.

They appear again and again—not just in scriptures, but in the sky, symbolically linked to the seven stars of the **Ursa Major** constellation. In this sense, they are not only mythological; they are **astronomical**, **philosophical**, and **symbolic**—beacons of eternal truth visible in both the heavens and the human soul.

More Than History

The Saptarishi belong to a time before linear time—a **mythic age** where the boundaries between gods and humans were fluid, where wisdom was received in silence, and where knowledge flowed like rivers from the Source.

In the Puranas, Mahabharata, Upanishads, and Vedas, they are referenced with deep reverence. Their names vary slightly depending on the cycle (Yuga) being described, but the most commonly recognized seven include:

1. **Vasistha**
2. **Atri**
3. **Bhrigu**
4. **Gautama**
5. **Jamadagni**
6. **Kashyapa**
7. **Bharadvaja** *(or sometimes Agastya or Vishvamitra, depending on tradition)*

Each of them represents a **facet of divine wisdom**—from discipline and devotion to cosmic insight, ancestral power, and sacred service.

Their stories often blur the lines between myth and memory, making them not merely characters of legend, but **cosmic archetypes** embedded in the spiritual DNA of India.

Carriers of the Vedas

According to ancient lore, the Saptarishi were responsible for **receiving and transmitting the Vedas**—not composing them, but **hearing** them (*śruti*) from the primordial vibration of the universe. In this sense, they were **receivers**, not authors—antennas tuned to the frequency of truth.

Their insights formed the foundation of **vedic knowledge, ritual, metaphysics, and cosmic law**. They were not priests in the formal sense—they were **seers**, perceiving the structure behind the structure, and handing it down not as dogma, but as **living alignment**.

Cosmic Cycles and Yugas

The idea of the Saptarishi is not frozen in a single age. In traditional cosmology, time moves in **vast cycles**, and in each era (*manvantara*), a new set of seven sages arises to guide humanity.

This makes the Saptarishi not a historical list, but a **recurring constellation** of wisdom-keepers—those who appear when the world forgets, to restore remembrance.

They are linked to the **North Star**, to the axis of spiritual orientation. They stand as reminders that no matter how dark the age, the light of wisdom remains **available, stable, and real**.

Inner Archetypes

Beyond mythology, the Seven Sages can be seen as **aspects of inner development**—archetypes that live within each seeker. Each sage represents a different approach to spiritual growth:

- The sage of silence
- The sage of discipline
- The sage of devotion
- The sage of action
- The sage of balance
- The sage of cosmic vision
- The sage of transmission and legacy

In this sense, the Saptarishi are not just **figures to study**, but **dimensions of being** to awaken.

Their teachings are not instructions—they are **vibrational templates**, reminding us that the ancient truths still echo in the modern heart.

Why They Still Matter

In an era obsessed with novelty and speed, the Saptarishi offer a kind of **deep time**—a return to the slow, rooted, contemplative rhythm of cosmic alignment.

They ask us:

- What wisdom did we once carry that we have now forgotten?
- Can we remember the patterns that uphold the world?
- Can we live in a way that honors not just history, but eternity?

Their presence is a call to become **students of the eternal**, to hear again what was once heard in silence, and to carry forward the light that has never gone out.

Queries for Reflection

- Do ancient myths carry more than metaphor—can they hold memory?
- What does it mean to be part of a lineage that extends beyond time?
- Are there sages within you—guiding, waiting, ready to be reawakened?
- What patterns of wisdom need to be restored in the age you inhabit?

5.8.2
Vasistha: The Sage of Balance, Dharma, and Inner Sovereignty

Among the Seven Sages of India, **Vasistha** shines as a figure of profound clarity—at once a philosopher, a priest, a mystic, and a guide to kings. He represents **balance** in its highest form: not compromise, but a harmony that emerges from deep **inner sovereignty**. His life and teachings, preserved across Vedic hymns, epics, and spiritual texts, reveal the enduring power of **dharma**—not as law, but as alignment with cosmic truth.

Vasistha is often portrayed as a **royal sage**, a spiritual advisor to kings, particularly to **King Dasharatha**, father of **Rama** in the *Ramayana.* He appears not as a distant ascetic, but as a householder sage who stands fully within the world, yet radiates detachment, wisdom, and poise.

The Embodiment of Dharma

The Sanskrit word *dharma* can be translated in many ways—duty, righteousness, truth, or cosmic order. But in Vasistha's life, dharma is not an external code—it is an **inner compass**. He acts not from impulse or ideology, but from a quiet center of discernment, where personal integrity and universal principle meet.

He teaches that each being has a unique nature, and to live in accordance with that nature is the root of peace. Dharma is not about obedience—it is about **listening to the harmony within** and aligning one's life accordingly.

In the *Ramayana*, when dramatic events threaten to derail the kingdom's stability or a prince's inner balance, it is Vasistha who brings perspective—

not with commandments, but with **questions and stories**. He does not impose. He illumines.

His gift is not just knowledge—it is **wisdom applied to the moment**.

Householder and Sage

Unlike many spiritual figures who renounced the world, Vasistha was a **householder**. He had a wife, **Arundhati**, herself revered for her fidelity, brilliance, and spiritual power. Their marriage is remembered as an ideal of **equally shared devotion and responsibility**.

This image matters. It shows that the path to enlightenment is not limited to caves and isolation. Vasistha proves that one can live in society, raise a family, perform duties—and still be **deeply free**.

This model would later inspire the Sikh tradition, as well as many Indian spiritual teachers who saw no conflict between the sacred and the everyday.

The Sage Who Stands Firm

Throughout the epics, Vasistha is tested—not only by external conflict but by encounters with other powerful figures, including **Vishvamitra**, a rival sage whose ambition and temper contrast sharply with Vasistha's steadiness.

In one famous episode, Vishvamitra attempts to seize Vasistha's divine cow, **Kamadhenu**, a symbol of spiritual abundance and purity. When Vasistha resists, he does so **without anger or fear**. His calmness neutralizes the force of his opponent.

"Brahma's might is greater than Kshatriya's strength," he says, showing that **spiritual power grounded in dharma** can overcome even violent force.

Here, Vasistha becomes an archetype of **non-reactive wisdom**—unshaken by provocation, anchored in the eternal, clear in what must be preserved.

The Yoga Vasistha: Inner Sovereignty Revealed

One of the most profound texts attributed to him is the **Yoga Vasistha**, a spiritual dialogue between Vasistha and Prince Rama. In this extraordinary scripture, Vasistha offers teachings on the nature of **mind, illusion, liberation, and consciousness**.

Far from simple moral instruction, the Yoga Vasistha dives into **nondual awareness**, exploring how reality is shaped by thought, how suffering is born of attachment, and how liberation arises when the mind becomes still and spacious.

Vasistha tells Rama that the world is **a projection of mind**—not to be rejected, but to be understood. He offers stories within stories, parables of dreamers and awakened souls, always returning to the same theme: **freedom lies within**.

"The mind is the creator of the world.
The wise consider the mind to be all.
Control of the mind is the supreme wisdom."
—(Yoga Vasistha)

This vision places Vasistha not only among sages, but among **seers of consciousness** a timeless voice that resonates with Advaita Vedanta, Buddhist emptiness, and modern contemplative psychology.

A Model for Today

In a world pulled between activism and burnout, ambition and avoidance, Vasistha offers a third way: **centered, grounded, engaged clarity**. He does not withdraw from challenge—he meets it from stillness.

His life says: Be in the world, but not of it. Fulfill your role, but from awareness. Speak, act, love, and serve—but never forget the **quiet fire within**.

This is the sovereignty he embodies: not dominance over others, but mastery of the self. Not escape from life, but participation without confusion.

Vasistha reminds us that true power is **rooted in presence**. That truth is not noise, but clarity. And that the path to peace is not dramatic—it is **steady, luminous, and real**.

5.8.3
Atri and Anasuya: Purity, Devotion, and the Triple Flame

Among the seven ancient Rishis, **Atri** holds a place of quiet radiance—a sage known not for dramatic confrontations or fierce asceticism, but for his deep **devotion, clarity, and inner purity**. Yet his legacy cannot be told apart from that of his wife, **Anasuya**, whose presence is equally revered in Indian tradition. Together, they represent an archetype that transcends gender: the **unified strength of love and virtue**, embodied in both masculine and feminine form.

Where Vasistha speaks of dharma and cosmic alignment, Atri and Anasuya reveal the **power of devotion**—not as passive reverence, but as an **active force** capable of transforming even the gods.

The Sage Who Sees in Silence

Atri is remembered as a **seer of Vedic hymns**, a master of *tapas* (austerity) and inner stillness. The *Rigveda* includes verses attributed to him—praising the Divine as **light, order, and truth**. But his presence in later mythology expands far beyond scholarly contribution.

He is said to have performed such intense meditation that his mind became like a clear sky—**transparent enough to reflect divinity itself**. From this meditative fire, it is said, emerged a blazing energy that even the gods had to acknowledge.

In this way, Atri becomes an image of **purity through simplicity**—the kind of sage who doesn't seek power, but becomes powerful through **presence** alone.

Anasuya: The Essence of Grace

If Atri is the silent flame, Anasuya is its gentle glow.

Her name means **"free from envy"**—a state of heart so pure that even divine beings were humbled in her presence. In a world of rivalry and ego, she embodied **selfless love, hospitality, and unwavering faith**.

One of the most well-known stories about her tells of a time when the **Trimurti—Brahma, Vishnu, and Shiva—tested her purity**, disguising themselves as wandering sages and asking her to feed them while completely unclothed. Undisturbed, she agreed—and by the power of her inner virtue, **transformed the gods into infants**, bathing and feeding them as a mother would.

This moment is not about supernatural display. It is a **spiritual metaphor**: that true virtue disarms all power, that love dissolves pretense, and that humility holds more strength than fear ever could.

Anasuya's purity is not naivety. It is a **radiant intelligence of the heart**, capable of cutting through illusion without violence.

The Triple Flame

From their union was born **Dattatreya**—a mysterious figure revered as a sage, yogi, and incarnation of the **Trimurti**. In him, the energies of creation (Brahma), preservation (Vishnu), and dissolution (Shiva) are said to be unified.

Dattatreya is often depicted as a wandering ascetic with three heads and four dogs (representing the Vedas) walking beside him. He teaches that the **Guru can be found in all of life**—in nature, in animals, in experience.

This **Triple Flame**—Atri, Anasuya, and Dattatreya—symbolizes a higher synthesis. Not hierarchy, but harmony. Not fragmentation, but fusion. It points to a spirituality that transcends form while honoring it—a lineage where **wisdom, devotion, and power meet in stillness**.

The Alchemy of Relationship

What makes Atri and Anasuya remarkable is not just their individual qualities, but their **partnership**. In many spiritual traditions, the feminine is sidelined or hidden behind the masculine teacher. Here, Anasuya is not only visible—she is **equal in reverence and stature**.

Together, they show that **spiritual growth can be nourished through relationship**, not hindered by it. Their home was not just a domestic setting—it was an **ashram**, a sacred ground of learning, healing, and transmission.

In them, we are reminded that purity is not about avoidance, but about **authentic integration**—living in the world without being entangled by it.

A Mirror for the Seeker

Atri and Anasuya ask nothing of us but sincerity. They invite us to cultivate a heart **free from envy**, a mind **clear and patient**, and a presence that **uplifts without dominance**.

They show that austerity can be gentle. That silence can speak. That love, when grounded in truth, can tame the mightiest of forces.

And that sometimes, the greatest spiritual power is simply the ability to serve, nourish, and love—**without agenda, without fear, without self-importance**.

Queries for Reflection

- What does "purity" mean in a modern world—can it be more than avoidance?
- Can relationships become sacred vessels for growth, not obstacles to it?
- Where in your life can devotion become an active force of transformation?

- What if humility, grace, and care are not soft qualities—but the most potent powers of all?

5.8.4
Bhrigu and the Eternal Record: Time, Karma and Cosmic Imprints

Among the Seven Sages, Bhrigu is perhaps the most enigmatic—his presence stretching not only across the epics and scriptures of ancient India, but through the mystical corridors of time itself. Known as a seer, priest, astrologer, and visionary, Bhrigu is most deeply associated with a singular mystery: the idea that everything we are, were, or will be is written—not as fate, but as possibility. His legacy invites us to consider the relationship between time and truth, karma and destiny, and the hidden record that lives within all things.

The Sage of Seeing Beyond Time

Bhrigu is remembered as one of the earliest **recorders of cosmic knowledge**, a sage who looked not only at the stars, but **through them**. He is often depicted as a master of **Jyotisha**, the ancient science of Vedic astrology—not as prediction for amusement, but as a sacred map of the soul's journey across lifetimes.

According to legend, he compiled the **Bhrigu Samhita**—a vast collection of palm-leaf manuscripts said to contain the life destinies of countless souls. These records, it is said, were dictated while Bhrigu was in deep meditative communion with the **Akashic field**—the subtle realm where all actions, thoughts, and choices leave imprints.

In modern times, seekers still travel to Indian villages in hopes of finding **"their leaf"**—a document that knows their name, lineage, past, and future, as if time were an open book and consciousness its reader.

Whether one takes the Bhrigu Samhita literally or symbolically, it represents something profound: that our lives are not isolated events, but threads in a **cosmic weave**, shaped by causes beyond immediate memory.

Karma as Cosmic Echo

To understand Bhrigu is to engage with **karma**—not as punishment or reward, but as the law of resonance. Every action, thought, and word creates a ripple. These ripples may return in this life or another, not as retribution, but as **opportunity for awareness**.

Bhrigu's vision was not fatalistic. He did not reduce human experience to a formula. Rather, he revealed that beneath the apparent chaos of life lies **a pattern**, and that to read it rightly is to live with greater freedom and compassion.

Karma, in this sense, is not a sentence. It is **a mirror**—showing us what we have planted, so we might choose how to respond.

Bhrigu's insights into karma extend beyond the individual. He understood collective karma, ancestral karma, and the karma of places and times. He saw that we are never isolated. We inherit, we affect, and we participate in **interwoven fields of consequence**.

The Eye of the Seer

In one well-known story from the *Mahabharata*, Bhrigu is asked to determine which member of the Trimurti—**Brahma, Vishnu, or Shiva**—is most worthy of worship. He tests each deity by deliberately provoking them.

When he insults Brahma, Brahma responds with anger. When he disturbs Shiva in meditation, Shiva becomes enraged. But when he strikes Vishnu on the chest, Vishnu simply **smiles**, asks if he is hurt, and offers him comfort.

Bhrigu bows in awe. The test was not about hierarchy. It was about **humility, equanimity, and unconditional love**. Vishnu passed not through strength, but through stillness.

This story reveals a deeper truth: that the one who sees clearly must also see with **compassion**. That insight without kindness is incomplete. That the highest form of wisdom does not dominate—it dissolves separation.

The Subtle Body of the Cosmos

Bhrigu's connection to the cosmic record aligns with other mystical traditions: the **Akashic Records** in esoteric philosophy, the **Book of Life** in Abrahamic thought, the **collective unconscious** in depth psychology. Each points to the idea that consciousness is not limited to the brain or body—it is **embedded in the field of existence**.

In this view, time is not a line, but a spiral. Memory is not stored only in neurons, but in **space itself**. And the sage is one who can listen—not just with ears, but with soul—to what the universe remembers.

Bhrigu reminds us that our lives are **readable**—not in fixed detail, but in vibrational tone. And that by becoming still, sincere, and clear, we too can begin to read.

The Modern Invitation

In a world obsessed with speed and novelty, Bhrigu invites us to slow down and look deeper. To consider that our struggles are not meaningless. That our gifts are not accidents. That our choices echo across time.

He does not call us to superstition. He calls us to **subtlety**. To pattern recognition. To soul inquiry.

He asks: What is your life echoing? What field are you shaping? And what forgotten knowing waits in silence to be remembered?

Queries for Reflection

- Do you sense that your life has patterns—echoes from the past, seeds for the future?
- How might your understanding of karma shift if you saw it not as judgment, but as invitation?

- Can you imagine that the universe carries a memory—not only of actions, but of intention?
- What would it mean to live with awareness of the unseen record your life is writing?

5.8.5
Gautama and the Power of Austerity, Mind and Reconciliation

Among the great sages of ancient India, **Gautama** Rishi stands as a figure of sharp intellect, deep spiritual practice, and moral resilience. While less widely known in popular spiritual discourse than Vasistha or Bhrigu, Gautama's legacy carries immense significance. His name is attached to **both high ascetic practice and deep ethical reflection**, and his life bridges themes of **discipline, reconciliation, and the transformative power of the mind**.

Gautama's story unfolds in many layers—through the Vedas, the epics, and later traditions. He is credited as a **seer of Rigvedic hymns**, the founder of the **Nyaya school of logic**, and a moral guide whose personal trials offered insight into the challenges of anger, forgiveness, and the subtle workings of karma.

His legacy is not only one of intellectual sharpness or ascetic power—it is a reminder that even the wise must face their own **inner reckonings**, and that **humility is the true mark of a sage**.

The Sage of Logic and Inquiry

Gautama is traditionally associated with the **Nyaya school**, one of the six classical systems of Indian philosophy. Nyaya, meaning "method" or "rule," emphasizes the use of **logic, reasoning, and critical inquiry** as a path to truth.

For Gautama, spiritual realization was not born of blind belief. It required **discernment, questioning, and clarity of thought**. The mind, when properly trained, was not an obstacle to liberation—it was a **vehicle**.

The Nyaya Sutras attributed to him outline categories of valid knowledge (*pramāṇa*), tools for debate, and methods for dispelling illusion. This philosophical rigor parallels later traditions in Buddhist and Jain thought, and even echoes elements of Western rational inquiry.

But Gautama did not value logic for its own sake. He sought freedom from suffering, clarity of perception, and liberation from delusion. In this way, he reminds us that **reason and reverence can walk together**.

A Life of Austerity and Misjudgment

While Gautama is remembered for mental discipline, one of the most famous stories surrounding him deals not with philosophy, but with **emotion, reputation, and redemption**.

According to the *Ramayana* and various Puranic sources, Gautama lived with his wife **Ahalya**, a woman of great beauty and intelligence. In some versions of the tale, the god Indra, driven by desire, disguises himself as Gautama and deceives Ahalya into sleeping with him.

When Gautama discovers the betrayal, he is consumed with rage. In some tellings, he curses Ahalya to become a stone; in others, she is rendered invisible, or exiled into the forest. The variations are many, but the theme is clear: even great sages are not immune to the storms of the heart.

Yet the story does not end in condemnation. Later, when **Rama** comes to Gautama's ashram, his presence **redeems and awakens Ahalya**. She is restored—not by punishment, but by grace. Gautama, too, is said to be reconciled.

This tale, layered and symbolic, has been interpreted in many ways:

- As a warning against ego and lust
- As a metaphor for illusion and awakening
- As a teaching about forgiveness, karma, and the healing of broken trust

Whatever its historical reading, it reflects something universal: that even in the life of a sage, there are **moments of fallibility**, and that true wisdom lies not in perfection, but in the **capacity to return to clarity and compassion**.

Mind as Mirror

In both his logic and his life story, Gautama invites us to examine the **mind**—not as an enemy, but as an instrument.

He teaches that the mind, when clouded by desire or emotion, distorts reality. But when stilled and purified through reflection, inquiry, and right conduct, it becomes like a **mirror**—able to reflect truth without distortion.

This aligns Gautama not only with Hindu thought, but also with the later insights of **Buddha**, who shared his name. Indeed, the historical Buddha's father was said to be from the **Gautama lineage**, and the emphasis on mind, clarity, and liberation is a bridge between Vedic and Buddhist traditions.

The Power of Reconciliation

Ultimately, Gautama's life is a lesson in **reconciliation**—between logic and devotion, between the individual and community, between justice and compassion, between the mistakes we make and the truths we rediscover.

He shows that the path of the sage is not a straight line. It curves, breaks, repents, and repairs. It requires not only discipline, but **forgiveness**—of others, and of self.

In a world where anger spreads quickly, and reputations are shattered with little hope of restoration, Gautama reminds us that **healing is possible**, and that the mark of spiritual maturity is not certainty—but the willingness to see again, to understand more deeply, to act more wisely the next time.

Queries for Reflection

- How can reason and devotion work together in your own path?
- What stories or judgments are you still carrying—and are they ready to be released?

- Can the mind become a mirror when we treat it with respect and discipline?
- What does reconciliation mean when we have failed or been hurt—and are we willing to walk that path?

5.8.6
Jamadagni and Parashurama: Lineage, Justice and the Warrior's Shadow

---

In the pantheon of the Seven Sages, **Jamadagni** is a figure of paradox—an austere and righteous seer whose legacy is bound to one of the most complex avatars of Vishnu: his son, **Parashurama**, the warrior-sage. Together, they form a lineage that explores **justice and violence, devotion and wrath, purity and power**—not as contradictions, but as energies that must be integrated, understood, and ultimately transcended.

Their story does not offer easy answers. It holds up a mirror to the shadow aspects of dharma—those moments when duty collides with morality, when sacred action borders on vengeance, and when the flame of righteousness becomes too hot to hold.

In Jamadagni and Parashurama, we are invited to explore the **tensions between spiritual ideals and worldly realities**—and what it means to inherit both light and fire.

Jamadagni: The Sage of Fierce Principle

Jamadagni was known for his **intense austerity**, mastery of mantras, and unwavering commitment to dharma. He lived with his wife **Renuka** in a forest ashram, sustained by simplicity and sacred knowledge. His **cow, Kamadhenu**, was said to fulfill all desires, symbolizing the abundance that flows from alignment with truth.

But Jamadagni's story turns dark when the external world crashes into his hermitage. In various accounts, the king **Kartavirya Arjuna**, intoxicated with power, tries to seize Kamadhenu. In response, Parashurama defends the family, slays the king, and sets off a chain of retributive violence.

Yet long before this act of vengeance, Jamadagni's own actions raise questions. In a well-known legend, he becomes suspicious of Renuka's purity after a fleeting mental distraction and commands his sons to **kill her** to preserve family honor. All refuse—except Parashurama, who obeys. As a reward, Jamadagni offers him a boon, and Parashurama asks for his mother to be restored.

The story is deeply troubling. But like many ancient myths, it may be read less as a literal event and more as a **symbolic confrontation**—between rigid principle and evolving conscience, between inherited codes and emerging clarity.

Jamadagni's unwavering dharma is admirable—and also dangerous. He represents the **potential rigidity of the spiritual path** when love and discernment are lost. He reminds us that **law without compassion can become a blade**.

Parashurama: The Warrior-Sage

Parashurama, the sixth avatar of Vishnu, is unlike any other. He is a Brahmin by birth and a Kshatriya by action—a **hybrid being** who challenges both caste boundaries and spiritual archetypes.

After witnessing his father's murder by royal forces, Parashurama takes up his axe and vows to **destroy the Kshatriya class**, which he sees as corrupted and arrogant. In some accounts, he **wipes them out twenty-one times**, restoring balance through the sword rather than through prayer.

And yet, he is also a yogi, a teacher, and a sage who passes on martial and spiritual knowledge to others—including Rama and Karna in the Mahabharata. He lives beyond time, appearing at turning points in Indian mythology as a **custodian of sacred power**, unbound by age or limitation.

Parashurama embodies the **warrior's shadow**: rage as a response to injustice, destruction as a tool of purification. His story asks a difficult question: Can violence be righteous? Can wrath ever serve the Divine?

The answer is not simple. But his life offers a warning and a teaching: that even divine wrath must be **tempered by inner work**, and that true power lies not in vengeance, but in integration.

Lineage and Inheritance

Together, Jamadagni and Parashurama embody a **lineage of fire**—where the transmission of knowledge is wrapped in trial, and where devotion can wear many faces.

They show us that spiritual inheritance is not only about light. Sometimes we receive **wounds, anger, unfinished karma**, and the tools of transformation. What we do with them is what defines the next generation of sages.

Jamadagni's sternness gave Parashurama form. Parashurama's fury had to give way to wisdom. Both were incomplete alone. Together, they point toward a **whole and complex path**—not of perfection, but of ongoing inner reconciliation.

The Warrior Within

We all carry some version of Parashurama within us—the part that rises against injustice, that burns at betrayal, that wants to act decisively when the world feels wrong.

The teaching is not to suppress this energy, but to **channel it**. To move from reactive violence to conscious power. To allow dharma to include not just law, but love. Not just duty, but compassion.

Parashurama eventually lays down his axe—not because he is defeated, but because he is **transformed**.

He becomes a teacher, a guide, a reminder that even divine fire must return to the hearth.

Queries for Reflection

- What inherited beliefs, temperaments, or wounds are you being called to transform?
- Can righteousness become dangerous when it lacks humility?

- What does it mean to carry power with consciousness—and when is it right to lay the sword down?
- How do you integrate the warrior and the sage within yourself?

5.8.7
Kashyapa: The Ancestor Sage and the Seeds of Civilization

In the constellation of ancient seers, **Kashyapa** stands at the roots. He is often called the **father of beings**, the progenitor of many races—gods (*devas*), demons (*asuras*), celestial musicians (*gandharvas*), snakes (*nagas*), birds (*garudas*), and even human sages. In Vedic and Puranic lore, he is not only a wise teacher—he is a **symbol of origins**.

If the other rishis are mirrors of consciousness, Kashyapa is **soil**—the fertile ground where lineages take root, where ideas become institutions, and where the mythic family tree of Indian civilization first spreads its branches.

Yet his story is not just about ancestry. Kashyapa invites us to contemplate **creation, complexity, and coexistence**—what happens when life multiplies, diverges, and learns to live together in tension.

The Father of Many

According to the Puranas, Kashyapa was the son of **Marichi**, one of the mind-born sons of Brahma. He married many wives, most notably the daughters of the progenitor **Daksha**. Each of these unions gave rise to a different category of beings:

- From **Aditi**, came the **Adityas**—celestial deities associated with light, order, and the preservation of dharma.
- From **Diti**, came the **Daityas**—powerful beings often in opposition to the gods.
- From **Kadru**, emerged the **nagas**, or serpents—linked with both protection and hidden knowledge.

- From **Vinata**, came the mighty bird **Garuda**, mount of Vishnu and enemy of the nagas.

These offspring symbolize not just physical creatures but **forces of the cosmos**—light and shadow, ascent and descent, freedom and bondage. Kashyapa, therefore, becomes a **bridge** between polarities, holding the entire spectrum of existence within his lineage.

His household is not peaceful—it is **archetypally diverse**. His legacy is not simplicity—it is **evolutionary complexity**.

The Balance of Creation

Kashyapa's presence reminds us that creation is not neat. It births both cooperation and conflict. He does not favor one lineage over another. His gift is not control, but **containment**—an ability to be the origin point of many truths, even when they contradict each other.

This makes him a model not of command, but of **holding space**—for gods and demons, serpents and birds, teachers and rebels.

We live in a world that mirrors this inheritance. Our inner and outer lives are filled with contradictions, competing instincts, diverging values. Kashyapa teaches that the answer is not to eliminate complexity, but to **learn how to host it wisely**.

This is a deeper kind of spiritual maturity—not transcendence through escape, but coherence through **receptivity**.

Ancestral Vision

In many traditions, to become wise is to become aware of one's **ancestors**—not only biological, but spiritual. Kashyapa is a reminder that every culture, every wisdom path, every being has **roots**.

To honor Kashyapa is to honor:

- The long arc of transmission

- The lineages that shaped our perception
- The truths we've inherited—along with the blind spots
- The teachers, voices, and unseen threads that brought us here

He invites us to trace our spiritual ancestry—not to create division, but to see how **everything belongs**. Even the shadow. Even the serpent.

The Builder of Foundations

While other sages focused on philosophy, Kashyapa is remembered for **institutional legacy**. His name is associated with early forms of medicine, astrology, governance, and ethics. He appears not only in Vedic hymns but also in Ayurveda and other classical systems.

In this way, he is not just an origin point of beings, but of **civilization**—a builder, a systematizer, a transmitter of applied knowledge.

He is the sage who teaches us that **wisdom must be seeded into culture**—that spiritual insight must also become practical structure, rooted not just in contemplation but in contribution.

Holding the Many

Kashyapa's story may feel less personal than others—less fiery, less dramatic. But it is no less essential. He is the one who stands at the root, quietly allowing **diversity to unfold**, and trusting that even conflict can be part of a greater becoming.

He is the **cosmic elder**—not controlling, but observing; not retreating, but watching life multiply and mature.

His wisdom is not loud. It is **settled, enduring, patient**—the kind of wisdom that holds the fire, but doesn't fan it unnecessarily.

He asks us:

- Can we carry what we've inherited with grace?
- Can we make room for what we don't fully understand?
- Can we be elders in the making—not because we know everything, but because we can hold everything?

Queries for Reflection

- What lineages—spiritual, cultural, or emotional—live through you?
- Can you honor both the light and shadow aspects of your inheritance?
- How do you hold space for contradiction—within yourself, your community, your path?
- What wisdom is ready to be seeded into your daily life—not as a thought, but as a foundation?

5.8.8
The Saptarishi Within: Eternal Wisdom in the Age of Forgetting

---

The stories of the Seven Sages—Vasistha, Atri, Bhrigu, Gautama, Jamadagni, Parashurama, and Kashyapa—are more than ancient memories. They are **living symbols**. Archetypes. Fragments of cosmic intelligence seeded into human consciousness across time.

To encounter them is not merely to study history or mythology. It is to engage in **deep remembering**—of truths that never truly left us, but which grow quiet in times of noise, distraction, and forgetting.

And we are living, perhaps, in just such a time.

The sages, we are told, will return whenever dharma needs to be restored. But what if their return is not external? What if it is **inward**—through our choices, our clarity, our courage, and our capacity to listen?

What if the **Saptarishi live in you**?

Archetypes as Inner Guides

Each sage represents not only a historical figure, but a **quality of consciousness**:

- **Vasistha** is the inner anchor—clarity in chaos, stillness in responsibility.
- **Atri and Anasuya** are the flame of devotion—gentle, pure, and unshakable.
- **Bhrigu** is deep perception—the memory of the soul, echoing through time.
- **Gautama** is discernment—truth through reasoning, forgiveness after fire.

- **Jamadagni and Parashurama** are righteous power—strength that must evolve into wisdom.
- **Kashyapa** is integration—ancestral presence, patiently holding the tension of life's diversity.

These energies are not locked in ancient texts. They are **potentials within your own being**. You may feel one of them stir in you during a difficult conversation, a moral dilemma, a moment of silence, or an act of spontaneous generosity.

The path of the seeker is not to copy the sages. It is to **awaken what they represent**.

A Mirror for Our Time

The age we live in is one of **acceleration and amnesia**. Technologies move faster than wisdom. Information multiplies while insight often thins. Ritual is forgotten, silence is rare, and the hunger for soul is often buried beneath survival or spectacle.

But the sages have been here before. The stories tell us: they appear in **times of collapse and confusion**—not to judge, but to **remind**.

They remind us that:

- The mind can become still.
- Dharma can be rediscovered.
- Anger can evolve into protection.
- Knowledge can be rooted in devotion.
- Spirituality is not escape—it is embodiment.
- True wisdom includes contradiction—and holds it gently.

In this way, the Saptarishi are not bound to one culture or tradition. They are **timeless teachers** encoded into the human journey itself.

Becoming the Next Lineage

The ancient texts speak of **new sets of Rishis** appearing in each cycle of time. In every *Manvantara*, a fresh group is said to rise, carry, and transmit the sacred.

What if the next sages are not waiting to be born—but waiting to be remembered?

What if the lineage is not about being chosen, but about **choosing to listen**?

To walk as a modern sage is not to wear robes or live in forests. It is to **see clearly**, **speak truly**, **act from integrity**, and **serve without pride**.

It is to be a presence in your family, community, or generation who carries **depth in a time of distraction**, and **rootedness in a time of disorientation**.

It is not easy. But it is necessary.

The Quiet Return

The sages do not shout. They return in whispers, symbols, patterns, and memories. They return when someone chooses to pause before reacting. When a difficult truth is spoken with love. When an ancient teaching is lived quietly, humbly, and without fanfare.

They return when you stop running. When you remember who you are.

They return in stillness. In song. In service.

They return through you.

Final Queries for 5.8

- Which of the Seven Sages resonates most with your current season of life—and why?
- What aspects of ancient wisdom are asking to be remembered through your actions?
- How might you hold space for contradiction and complexity with grace?
- What if you didn't wait for the next sage to appear—because you are already being called to become one?

Glossary — Part 5: Avatars, Prophets, and Sacred Teachers

Ahimsa – A principle of non-violence found in Hinduism, Buddhism, and Jainism; central to the teachings of many spiritual figures, including the Buddha and Gandhi.

Anatman – In Buddhist philosophy, the doctrine of "non-self," suggesting that the idea of a permanent, unchanging soul is an illusion.

Avatar – A Sanskrit term meaning "descent"; refers to the incarnation of a divine being on Earth, most notably in the forms of Krishna and Rama in Hinduism.

Bhakti – A devotional path in Hinduism centered on loving devotion to a personal deity, such as Krishna or Rama.

Bodhisattva – In Mahayana Buddhism, an enlightened being who chooses to remain in the cycle of rebirth to help all sentient beings attain liberation.

Covenant – A sacred agreement or promise between God and a person or people, especially significant in the Hebrew Bible and the life of Moses.

Dharma – A complex term in Indian religions meaning cosmic law, moral order, or sacred duty; central to the teachings of Krishna, the Buddha, and many others.

Enlightenment (Bodhi) – The state of awakening to the true nature of reality, as attained by the Buddha under the Bodhi tree.

Gabriel (Jibril) – The archangel who delivered the revelations of the Qur'an to the Prophet Muhammad; also appears in Jewish and Christian texts.

Hadith – Collections of sayings, actions, and approvals of the Prophet Muhammad that supplement the Qur'an in Islamic tradition.

Immanence – The belief that the Divine is present and active within the material world and human experience.

Kenosis – A Christian theological term meaning "self-emptying," used to describe Christ's humility and surrender of divine privileges in becoming human.

Law (Torah / Sharia) – In Judaism, the Torah represents divine law revealed to Moses. In Islam, Sharia is the divinely ordained path, including legal, ethical, and spiritual guidance.

Logos – A Greek philosophical term meaning "Word" or "Reason"; in Christianity, it refers to Christ as the divine Word made flesh.

Meditation (Dhyana) – A disciplined spiritual practice found across traditions, used by the Buddha and other sages to cultivate awareness, clarity, and union with the Divine.

Messiah – A prophesied anointed one in Judaism and Christianity; Christians believe Jesus fulfills this role, while others await a future deliverer.

Monotheism – The belief in one singular, all-powerful God, central to Judaism, Christianity, and Islam.

Nabi – Arabic for prophet; a term used in Islam to denote those sent by God to guide humanity, including Moses, Jesus, and Muhammad.

Nirvana – The ultimate goal in Buddhism: liberation from suffering and the cycle of rebirth.

Prophet – A person chosen to convey divine messages to humanity; central figures in Abrahamic traditions and honored in others as spiritual seers or voices of truth.

Qur'an – The holy scripture of Islam, believed to be the direct word of God revealed to the Prophet Muhammad.

Revelation – A disclosure of divine truth, often through scripture, visions, or messengers; foundational to the roles of prophets across traditions.

Rigveda – An ancient Vedic text forming part of the foundational scriptures of Hinduism; predates the epics involving Krishna and Rama.

Sangha – The Buddhist community of monks, nuns, and laypeople dedicated to following the teachings of the Buddha.

Seal of the Prophets – A title for Muhammad in Islamic theology, signifying that he is the final prophet in a long line of messengers.

Shekinah – A Hebrew term for the Divine Presence; associated with the indwelling glory of God, particularly in Jewish mystical and prophetic traditions.

Shunyata – The Buddhist concept of "emptiness," indicating the lack of inherent existence in all things; a profound realization on the path to enlightenment.

Sutra – A sacred text or scripture in Buddhism and Hinduism, often containing teachings attributed to great sages or deities.

Tao (Dao) – The ineffable Way or Path in Taoism, echoed in the teachings of Laozi and reflected in Jesus and other sages who embodied spiritual flow.

Tikkun Olam – A Hebrew phrase meaning "repair of the world"; reflects the prophetic Jewish mission to bring justice, compassion, and wholeness to society.

Ummah – The global community of Muslims, united by faith and the teachings of the Prophet Muhammad.

… ✧ …

www.ingramcontent.com/pod-product-compliance
Lightning Source LLC
LaVergne TN
LVHW022210200926
848481LV00006B/1912
* 9 7 9 8 2 1 8 7 5 6 0 4 8 *